PIGEON GUIDED MISSILES

PIGEON GUIDED MISSILES

AND 49 OTHER IDEAS THAT NEVER TOOK OFF

JAMES MOORE & PAUL NERO

First published 2011

The History Press
The Mill, Brimscombe Port
Stroud, Gloucestershire, GL5 2QG
www.thehistorypress.co.uk

British Library Cataloguing in Publication Data.
A catalogue record for this book is available from the British Library.

ISBN 978 0 7524 5990 5

Typesetting and origination by The History Press
Printed in EU for The History Press.

CONTENTS

OUR THANKS

Like all writers, we earn our livelihoods not just through the words we put on the page, but on ideas; some big, some small, a few exciting, others more mundane. Occasionally we go through a purple patch of rather splendid ideas; more frequently our concepts fall by the wayside, perhaps even laughed at by our clients and families and friends. Or worse still, ignored. That's when we know we've really got things wrong. Now if we were adept at coming up with questionable ideas, surely it must happen to others? With this germ of a concept, the origins of Pigeon-Guided Missiles were formed. But we couldn't have done it alone. For making sense of some subjects in which we were hardly experts, for sourcing pictures where we had none, and for the occasional sanity or fact check, we needed help.

So we'd like to extend our warmest thanks to: Linda Bailey, Malcolm Barres-Baker, Tim Clarke, Jason Flahardy, Professor Christian G. Fritz, Amy Frost, Harry Gates, Stephen Guy, Simon Hamlet, Jan Hebditch, Richard Hebditch, Patricia Hansen, Ted Huetter, Aideen Jenkins, Owen Jones, Matthew Marshall, Amanda McNally, Geoff Moore, Laurie Moore, Philippa Moore, Sam Moore, Tamsin Moore, Dr Tom Moore, Marguerite Moran, Dr Claire Nesbitt, Neil Paterson, William Poole, Hannah Reyonlds, Kasey Richter, Sarah Sarkhel, David Smith, Robert Smith, Peter Spurgeon, Mia Sykes, Julie S. Vargas, Spencer Vignes, Paul Waddington and Dean Weber.

Thank you all.

INTRODUCTION
IN SEARCH OF HISTORY'S LOST IDEAS

If at first the idea is not absurd, then there is no hope for it.
Albert Einstein

Success is going from failure to failure without losing enthusiasm.
Winston Churchill

The past is littered with examples of grandiose schemes and ambitious ideas that never quite took off. While there are plenty of books about the visions, plans and inventions that did go on to transform our world, these heroic 'might have beens' have found themselves largely forgotten, carelessly dumped on the scrapheap of history.

This book sets out to rescue some of those incredible concepts and dreams, which, however briefly, promised to change our lives and the face of our planet. It also reveals the fascinating, flawed and tirelessly optimistic characters behind them. From soaring edifices that never were, to fanciful devices to change our daily lives; some of the ideas are simple, others breathtakingly outlandish. Eccentric engines of war, peculiar methods of transport, sporting follies and nation-building blunders – they're all here. Each example proves the precarious nature of success and shows how, but for a bit of serendipity, the world we live in today could have been very different. They are also testament to the dedication, inspiration and, at times, sheer bloody-mindedness of the people who conceived them.

Pigeon-Guided Missiles: and 49 Other Ideas that Never Took Off covers a lot of ground and we don't pretend to have been scientific in our choices. But in choosing these stories, we've relied on a few guiding principles. Every chapter endeavours to reveal a relatively unknown proposition from history, investigates what drove its proponents and why they failed. We have looked for engaging tales that are often humorous and sometimes tragic, but always contain the seeds of truly radical thinking.

In this book we'll discover that some, like Sir Edward Watkin's attempt to build a rival to the Eiffel Tower in London, or William Beckford's enormous gothic home, were the victim of bad planning. Others, such as Fulton's flying car, succumbed to lack of cash. The pyramid in London's Trafalgar Square and Bessemer's ship to cure seasickness were simply too audacious, while the first gas-powered traffic lights were too ahead of their time. Radar-style warning dishes made from concrete were overtaken by technological developments. Robert Fludd's perpetual motion machine and Harry Grindell Matthews' death ray simply seemed to defy the laws of physics. A lot, like so many good ideas, were just unlucky; failing to catch the eye of those who mattered, dismissed as preposterous or left gathering dust thanks to the prevailing political or commercial climate.

One thing becomes very clear from our research – that history is so often driven by a few industrious men and women, who, verging on the obsessive, never stop coming up with ideas that they genuinely believe will be a step forward for science, will help mankind or enhance the world we live in. As this book shows, they don't always get it right, but the story of how they fail, often spectacularly, is endlessly captivating and there is usually something to be salvaged from the ashes of their efforts – even if it is occasionally just a good belly laugh.

1

PIGEON-GUIDED MISSILES

Since Hannibal crossed the Alps with his elephants in the third century BC, aiming to conquer Rome, mankind has frequently used all sorts of animals as tools of war.

One creature to make a considerable contribution to the sphere of human conflict is the humble pigeon. During the Second World War some 250,000 homing pigeons served with British forces. Thirty-two were even awarded the Dickin Medal, the military's version of the VC for animals. The US had its own Pigeon Service and one of its ranks, nicknamed GI Joe, was credited with saving 1,000 lives.

Yet some believed the pigeon's military capabilities lay beyond carrying messages. In 1945 an official at Britain's Air Ministry Pigeon Section, Lea Rayner, spoke of how pigeons might carry explosives and even become vehicles for bacterial warfare. But it was in America that the boldest designs for the use of pigeons against the enemy were formulated – by the renowned behavioural scientist B.F. Skinner. Through his work with the birds, Skinner believed that they held the key to perfecting the next step in military technology, the guided missile.

Burrhus Frederic Skinner, born in 1904, was no quack; and for a time the US authorities didn't think his ideas were bird brained either. They took him very seriously indeed and

The nose cone of a prototype pigeon-guided missile, as pioneered by behavioural psychologist B.F.Skinner in the 1940s.

funded his pigeon project, hoping that harnessing the abilities of these winged wonders would give them the critical edge in defeating the Axis powers, and the key to their country's future defence.

By the time the Second World War broke out, Skinner, who later went on to become a professor at Harvard, was already a successful psychologist. He had helped pioneer a theory called operant conditioning. In essence he believed that animal and human behaviour was reinforced by external factors like rewards or punishments. Legend has it that, after seeing a flock of birds flying alongside a train, Skinner suddenly realised that his research on conditioning had a practical application for the war effort. He could train birds to guide missiles to their targets. 'It was no longer merely an experimental analysis. It had given rise to a technology,' he said.

In 1940 his initial work at the University of Minnesota had shown that pigeons could be trained to repeatedly pick out a target by using the reinforcement technique of pecking at pieces of grain. In this way they could be conditioned to keep pecking at a target and their movements linked to mechanisms that would alter the guidance controls of the missiles. One of Skinner's birds pecked at an image more than 10,000 times in forty-five minutes and, by 1941, he was able to show that it was possible to train pigeons to steer towards small model ships. Despite this, after taking his research to defence officials, he was told that his proposal 'did not warrant further development at this time'.

Then, in 1942, Skinner's work was suddenly dusted off by researchers looking for a psychologist to train dogs to steer anti-submarine torpedoes. It eventually led to a grant of $25,000 from the US Government to develop the idea via a company called General Mills, who also wanted to do their bit for the war effort. During this further research, Skinner found that a pigeon would track an object by pecking at a screen even under all sorts of difficult conditions, including rapid descent and the noise of explosions. Subsequently, plans were drawn up to experiment using the pigeons inside the new, aptly named, 'pelican' missiles.

Skinner's plan was to load three pigeons into their own pressurised chambers inside the missile nose cone. Lenses in the missile threw up an image of the target on a glass screen. Once they saw it the pigeons would start to peck and their movements translated into adjustments in the missile's guidance rudders. The results amazed fellow scientists with some saying that the pigeons' accuracy rivalled that achieved by radar. In addition, using pigeons in this capacity didn't involve radio signals that could be jammed by the enemy.

Skinner himself explained why three birds were needed:

> When a missile is falling toward two ships at sea, for example, there is no guarantee that all three pigeons will steer toward the same ship. But at least two must agree, and the third can then be punished for his minority opinion. Under proper contingencies of reinforcement a punished bird will shift immediately to the majority view. When all three are

working on one ship, any defection is immediately punished and corrected.

Prototype missiles were built and legions of pigeons prepared
to be trained up for service. Mass production of the missiles
was primed to snap into action in just thirty days. But, at a
high level demonstration in 1944, government officials simply
couldn't get their heads round the idea that pigeons could
ever be satisfactorily controlled. Skinner said: 'The spectacle
of a living pigeon carrying out its assignment, no matter how
beautifully, simply reminded the committee of how utterly
fantastic our proposal was.'

That wasn't quite the end of the project. In 1948 it was
revived by the US Navy under the name Project Orcon
(organic control). This time a pigeon's beak was fitted with
a gold electrode which would hit a semi-conductive plate to
report the target's position to the missile's controlling mechanism. But a line was finally drawn under the project in 1953
when further developments in electronic guidance systems
made the pigeons redundant.

Skinner's pigeon-guided missile system never took off.
A plan for bat bombs, also tested during the 1940s, met a similar fate. The creatures were to be fitted with mini parachutes
and timed incendiary devices and released en masse from aircraft. The idea was that, after being dropped over enemy cities,
they would naturally look for places to roost in buildings and
set them alight, causing a firestorm. The $2 million project was
abandoned not long after a colony of bats accidentally blew
up a fuel tank at the Carlsbad Air Force base in New Mexico.
Meanwhile, Soviet forces trained dogs to blow up German
tanks. This produced mixed results when some of the dogs ran
towards their own lines: they'd been trained to listen for the
engine noises of Soviet tanks, not enemy German ones.

The world's military did not give up on using animals in
warfare in the ensuing decades, however. More recently the
Soviets experimented on training so-called kamikaze dolphins
at a secret base on the Pacific coast, to see if they could carry

mines to attack enemy warships. The programme was halted by the break up of the Soviet Union and some of the kamikaze dolphins were even sold to the Iranians, though quite what they did with them remains unclear.

Russians have trained seals to locate mines with more success, while the US today uses sea lions to protect one of its important naval bases against terrorists. The sea lions have been trained to place a cuff, like a handcuff, on enemy divers, who can then be reeled to nearby boats. Animals in military service also still save lives on a daily basis; in 2010 a black Labrador called Treo was awarded the Dickin Medal for his work sniffing out bombs in Afghanistan.

Well after the Second World War MI5 kept a stock of trained pigeons for its security work. And of his experiments with pigeon-based warfare, B.F. Skinner himself admitted: 'I knew that in the eyes of the world we were crazy.' But, in an article in American Psychologist in 1960, he also told how he believed that many genuine scientific advances would never have been achieved without the odd 'crackpot' idea. His research on pigeon-guided missiles certainly demonstrated how powerful the idea of conditioning could be. Incredibly, six years after the first pigeon project had ended, Skinner found that some of the pigeons he had trained could still identify the same targets.

THE INTERNATIONAL
'HOT AIR' AIRLINE

It was to be the first international airline and the first airmail service. But the difficulties of launching the Aerial Transit Company, which planned to shrink the globe from its head-quarters at a lace factory in Somerset, proved insurmountable. This was 1842, the age of steam, and getting an aircraft off the earth's surface was proving to be a lot of hot air.

Unlike the story of Icarus, who attempted to flee Crete by flapping feathered wings designed by his dad Daedalus, the international steam-powered airline is no myth. Inventors and entrepreneurs William Samuel Henson and John Stringfellow followed in a succession of aspiring aeronautical engineers who, since Icarus, have tried to defy gravity with an array of hapless devices. In 1540 in Portugal, João Torto believed he'd bettered Daedalus' design by using two pairs of cloth wings while wearing a helmet in the shape of an eagle's head, but this only served to double the speed with which he plummeted to earth after jumping from a cathedral tower. In 1712 Frenchman Charles Allard strapped on wings of an improved design and launched himself from the Terrasse de St Germain in the direction of Bois du Vésinet, but only completed the Terrasse de St Germain part of the journey before dying from multiple injuries. And in 1739 steeplejack and occasional tightrope walker Robert Cadman entertained the residents of Shrewsbury with his own spectacular and rather messy death, while attempting to soar from the spire of St Mary's Church to the far bank of the River Severn with the help, or hindrance as it turns out, of an overly taut piece of cord that snapped under his weight. This may have been more an extravagant abseil than genuine flight, but the result was much the same. Cadman's wife, passing round a hat below, dropped all the donations when told how he had been 'dashed to pieces' while her back was turned. A stone memorial now commemorates the achievement, recalling Cadman's:

> attempt to fly from this high spire
> across the Sabrine he did acquire
> His fatal end.

So the fact that no one had actually been airborne for more than a few seconds – and only then in a strictly vertical and downwards trajectory at speeds they hadn't expected – should have made the task of starting the world's first international airline rather daunting. But in the industrial age, along with the rapid advancements of scientific theory, inventions were coming thick and fast. There was no reason that man shouldn't get into the air and return to earth thousands of miles from his starting point with his limbs attached in the traditional formation. Indeed Henson, the son of a lace factory owner, and Stringfellow, a toolmaker who made bobbins, thought they knew how.

With the expertise of their friend Sir George Cayley, who had designed the first glider to carry a human being and who is sometimes described as the father of aviation, they calculated what it would take to build a passenger-carrying, self-propelled aircraft. Having studied birds in flight extensively and – this being the heyday of natural history – practiced their trajectory across a room with stuffed ones, they felt they were comfortable with the concepts of movement in three dimensions. 'My invention,' said Henson in his 1842 patent, 'will have the same relation to the general machine which the extended wings of a bird have to the body when a bird is skimming in the air.' Such aeronautical taxidermy was all very well, but they also needed mechanics. A boiler, a paddle wheel, and somewhere to sit while the pilot stoked the engine would suit perfectly.

Before beginning international flight operations, they first had to build an aircraft. Henson's designs were elegant and Stringfellow's lightweight engines, in deft contrast to the other industrial monsters of the age, inspired. Under prevailing UK patent rules that didn't require evidence that inventions worked, they would create a machine 'to convey letters, goods and passengers from place to place through the air.' They called it the Aeriel or the Aeriel Steam Carriage; a monoplane with a wingspan of 150ft that would carry a dozen passengers 1,000 miles, although with a top speed of just 50 mph, this would mean 20 airborne hours, which was asking a lot for a lightweight engine of 50hp (37 kW). In terms of practicalities, 2 square feet of supporting surface added a pound of weight, requiring the engine to generate 20hp per ton to stay airborne. It worked on paper.

With a patent granted and investment raised from a small group of friends, attention turned to the second key aspect of the plan: publicity. In this, Henson and Stringfellow went to town. If you're going to launch the world's first international airline, then the world has to know about it. To Sir George Cayley's consternation – he was to withdraw eventually from investment – posters and flyers began to appear of the plane in the unlikeliest locations: here's the Aeriel in flight over London; and here it is again, this time over the Giza pyramids; and now India. An advertisement of the plane soaring over China was a particular favourite of the two inventors, who hoped to build eastern markets for British commerce.

China, though, would have to wait, for Chard had to be conquered first. Residents of the Somerset town were to be Henson and Stringfelllow's first audience – and, it was to be hoped, future airfare paying customers. The Ariel might not have been able to carry passengers at this – nor indeed any –

stage, but the prototype was impressive enough: the first plane of modern construction with a 12ft wingspan, three-wheeled landing gear and power from two contra-rotating bladed propellers. Its first run, in contrast, was a thunderous disappointment all round. After shuddering down a short ramp, instead of taking to the air, it came to a pitiful stop.

Undaunted, Henson and Stringfellow set to work on a bigger model, with a broader 20ft wingspan and a steam engine that delivered greater power. Marketing was also stepped up. Promotional handkerchiefs, trays, wall tapestries and lace placemats joined the posters and newspaper advertisements. It was nothing but hot air. Over the course of three years from 1844, the larger model plane was tested over and over and over again. It never flew. Finally, at his wit's end, Henson took drastic action; he married his girlfriend and, packing up the whole enterprise as a rum job, immigrated to the States, presumably by steamer. Stringfellow soldiered on, his aeronautical ambitions higher than ever. And for him at least, success followed.

In a redundant lace mill in June 1848, the model plane was launched from an inclined wire. Stringfellow had flight! Short flight, but flight nevertheless. According to some reports, the plane flew straight for about 30ft (although other reports dismiss its achievements as nothing more than 'a short hop'). Delighted whatever the distance, Stringfellow repeated the exercise and said the plane even sometimes gained altitude – which would be handy for the journey to China. Accolades followed. When the Aeronautical Society awarded Stringfellow a £100 prize, Scientific American magazine was impressed. He was, they claimed, responsible for 'probably the lightest ever steam engine ever constructed'.

By 1869, Stringfellow had developed the engine further. It now turned 3,000 revolutions a minute and, reported Scientific American, just 'three minutes after lighting the fire the pressure was up to 30 pounds and in seven minutes the full working pressure of 100 pounds, generating just over one horsepower'. It was laughably insufficient to get passengers to the pyramids or New Zealand. It wouldn't even get a full size plane airborne at all. But Henson and Stringfellow's invention did become the first powered plane in the world to fly. While only a model, it was to form a landmark on the way to the first passenger flights. Just before his death in 1883, Stringfellow said: 'Somebody must do better than I before we succeed with aerial navigations.' Just twenty years later, with the internal combustion engine overtaking steam as the power of choice, the Wright Brothers were in the air.

THE 'SPRUCE GOOSE'

Howard Hughes surely defined the word eccentric. One of the richest men in the world, he was famous for his maverick movie making, addiction to drugs and love of beautiful actresses. Stories of the billionaire's bizarre behaviour are legion too. Legend has it that he once gave staff precise instructions on how to lift a toilet seat. On another occasion, he became obsessed with designing a complicated cantilevered bra for one of the stars in his movie *The Outlaw*. Hughes was also an accomplished aviator and innovative aircraft manufacturer who was, for a time, considered a crucial cog in visionary plans to defeat Hitler.

The early twentieth century gave birth to many unsuccessful aircraft. There was the Bristol Brabazon, for example, the trans-Atlantic flop of the 1940s which was just too big, too expensive and too luxurious for the job. But none of the many flying failures of the century had quite the glamour and audaciousness of Howard Hughes' H-4 Hercules, better known as the 'Spruce Goose'. In its day it was the world's largest plane, and even today still holds the record for the longest wingspan of any aircraft. It was an invention that had all the grandeur, eccentricity and capacity for controversy as the man himself.

Despite his famous foibles, Hughes, born in 1905 to an oil-industry entrepreneur, was an undoubtedly clever man. At 14 he took his first flying lesson and it led to a lifelong love of flight. Later, in 1930, he would splash out a then-vast sum of $4 million to make a First World War flying film called *Hell's Angels*. In 1932 he set up his own aviation company, Hughes Aircraft, and went on to help design a number of planes, win flying trophies and set new air speed records. In 1938 he even flew right round the world breaking Charles Lindbergh's New York to Paris record in the process.

When the war came along Hughes' aircraft company was only employing a handful of people. By the end of the Second World War it would be employing tens of thousands, and along with aircraft development his subsidiaries would go on to supply the US military with everything from radar to air-to-air missiles and on board fire control systems. In 1942 he was

Howard Hughes' H-4 Hercules on its maiden (and only) flight in 1947. Dubbed the 'Spruce Goose', it was the world's largest plane and still holds the record for the longest wingspan.

approached by the industrialist and shipbuilder Henry Kaiser, who had a plan to tackle the huge losses being suffered by US shipping at the hands of German U-boats. A massive 800,000 tons of shipping had been sunk and Kaiser felt that the way round this was to build a huge aeroplane, capable of carrying 750 troops, and even tanks, across the Atlantic. Hughes enthusiastically embraced Kaiser's idea and together they managed to obtain $18 million in funds to build a huge flying boat to be called the HK-1. According to later reports, President Roosevelt himself overruled experts to give Hughes the contract, along with another for a long-range reconnaissance plane, the XF-11.

The pair were told to build three HK-1 prototypes to be ready within two years, and work duly began building the plane on a site in Southern California. Like many of Hughes' films, the final design was truly epic. The wingspan of the Spruce Goose, as it was nicknamed by a newspaperman of the day, was 320ft. To put that in a modern day context, the Airbus A380's wingspan is a mere 261ft, and a Boeing 747's just 211ft. The wings of the Goose were 13ft thick and the tail-fin alone was the height of an eight-storey building. The beast weighed 400,000lbs and its propellers were more than 27ft long. Together the plane's eight 5,000lb, 28 cylinder engines boasted 24,000hp, while the plane could carry 14,000 gallons of fuel. All the more remarkable, because of wartime restrictions, it had to be made out of wood rather than metal – hence its moniker. In fact the Spruce Goose was mostly made out of birch plywood, not spruce.

But all the time needed to carry out research on such an enormous plane, as well as supply problems and Hughes' debilitating perfectionism, meant that construction dragged; and in 1944 Kaiser quit the project, sensing the way the wind was blowing. The end of the war came and went and the aircraft, now officially called the H-4, still hadn't made it to its dockside hanger. But Hughes didn't care; he shovelled another $7 million of his own money into the plane's manufacture despite the fact that, with the conflict over, the original need for such an aircraft had gone. Then, in summer 1946, Hughes suffered terrible injuries when, piloting his other experimental plane for the government, the XF-11, he crashed into a housing estate. That wasn't his only difficulty. The government had started to ask questions. Where exactly, it wanted to know, had all its money gone? Why had no aircraft been delivered in four years?

In 1947 the US Senate War Investigating Committee began its hearings into Hughes' company, amid lewd allegations of a history of sleazy backhanders involving Hollywood starlets. When Hughes appeared in front of the committee he brazened it out saying:

The Hercules was a monumental undertaking. It is the largest aircraft ever built. It is over five stories tall with a wingspan longer than a football field. That's more than a city block. Now, I put the sweat of my life into this thing. I have my reputation all rolled up in it and I have stated several times that if it's a failure I'll probably leave this country and never come back. And I mean it!

But Hughes had been seriously stung by critics who described his beloved creation as a 'flying lumberyard' and taunted him that it would never fly. He loathed its derogatory Spruce Goose nickname. Hughes decided to show them all. During the hearings he had the only Goose ever completed at last delivered to Long Beach, California, and readied the craft for its first flight.

On 2 November, with Hughes himself at the controls, the plane skimmed across the water before suddenly gliding into the air for the first time. Hughes flew it at just 70ft for a total of 1 mile. It wasn't much, but Hughes had proved that the Spruce Goose could, indeed, fly. The investigation into Hughes' affairs never really got anywhere, but the government still cancelled its order for the plane.

Even by 1948 it was clear that Hughes, an obsessive, hadn't given up. A glowing report in *Popular Mechanics* reckoned the plane 'is slated to be for many years the biggest plane that ever flew'. Sadly it was also destined to be the biggest plane that never flew again. Still Hughes, who subsequently became a recluse, could never let go. For the next quarter of a century he kept it in a specially designed, climate controlled hanger at a cost of $1 million a year.

After his death in 1976 no one quite seemed to know what to do with it. Even Disney didn't want it. Finally, in the 1990s, the Goose found a buyer who wanted to make it part of his own collection of wartime aircraft. And, in 2001, the fully restored dinosaur went on display at its new home, the Evergreen Aviation Museum in Oregon. There the plane remains as a fitting memorial to a man who just didn't know when to stop.

4

A SOUND PLAN
FOR DEFENCE

Just beyond a row of unremarkable looking bungalows, near the lonely foreland of Dungeness on Britain's south coast, a group of strange concrete monoliths loom out of a flat gravel landscape. Today these crumbling concave edifices provide an eerie spectacle in the hush of their marshy surroundings. They are, in fact, the bizarre remains of a cutting-edge defence system designed to protect the country from the roar of war.

Before radar famously helped Britain fend off the menace posed by the Luftwaffe in 1940, government technicians had been working on another top-secret technology to protect the country from airborne attack. All that is now left as a reminder of this fascinating chapter in the defence of the realm are a few imposing ruins. These huge acoustic mirrors, abandoned in the Kent countryside, were once the key to the plan.

The strange story of the sound mirrors begins amid the carnage of the trenches. In 1915, as the First World War raged, a bright solider with a degree in physics called William Sansome Tucker was working with a unit experimenting on sound ranging – a way to detect the location of enemy artillery from the sound waves produced by shell fire. Tucker managed to come up with a 'hot wire' microphone which eventually allowed units to pinpoint artillery to about 50m,

Acoustic early-warning mirrors near Dungeness on Britain's south coast. Constructed from concrete in the 1920s, they were designed to detect enemy aircraft out to sea.

allowing more accurate retaliatory attacks. Meanwhile, back in Britain, experiments with sound mirrors were already afoot. The Zeppelin threat to Britain had led a Professor Mather to start looking into huge dishes which would focus the sound of enemy aircraft and airships so that they could be heard, though still well out to sea, before they came into visual range. In 1916 he'd had a crude 16ft mirror dug into a chalk hillside in Kent, and went on to do early work with concrete versions of these sound mirrors. One of these early designs was able to pick up a plane 10 miles away.

After the war, Tucker got the job of Director of Acoustical Research at Air Defense Experimental Establishment, where he was able to bring together his work on microphones and the sound enhancing capability of the satellite-dish-style reflectors. His efforts were centered on a site near Dungeness. This location, just yards from the coast, was felt to be far enough away from extraneous noise, and in 1928 Tucker oversaw the construction of a concrete mirror, 20ft in diameter. The general idea was that the low frequency sound waves from aircraft would be harnessed and concentrated, then picked up by a microphone. A man would stand next to the mirror listening in with a stethoscope and, in this way, be able to give the alert when aircraft were approaching. A bigger, 30ft mirror was built in 1930. Its larger surface area made it more accurate, and the 'listener' was able to work from inside a sound-proofed compartment. By moving the microphone collector horizontally and vertically using foot pedals and a wheel, he could identify the direction the sound was coming from and get a bearing on, say, a squadron of incoming aircraft.

In the same year another huge mirror with a different style of design was built – 200ft long and 26ft high with a curve of 150°. In this design a set of twenty microphones were built in to help pick up the sounds. The mirror's new design also enabled it to pick up sounds with the longer wavelengths usually created by aircraft. In 1932 this mirror was able, on one occasion, to locate aircraft 20 miles away. The human ear could only pick them up around 6 miles away. Tucker was beginning to finesse the technology.

For all the improvements in his designs, however, the mirrors were still hampered by the fact that sound waves travel relatively slowly. By the time the aircraft had been plotted it was already very close. In the years that Tucker had been working on the project aircraft speed had also been gradually increasing, making the sound mirrors less effective. There might, for instance, be only around four minutes' warning before the identified aircraft were overhead. Even at the lonely spot he'd chosen for the experiments the noise of wind, ships at sea and even car traffic could interfere with the results too.

Tucker nevertheless carried on. Perhaps the words of MP Stanley Baldwin that 'the bomber will always get through' were echoing in his ears. He recommended a line of mirrors should be built along the coast from The Wash to Dorset, and, in 1934, the government made preparations to construct the first salvo of these defensive mirrors around the Thames Estuary. There was to be a string of 200ft mirrors at 16 mile intervals with smaller 30ft-style mirrors filling the gaps.

In 1935 came news that was to make Tucker's work look like an expensive white elephant. While Tucker had been slaving away on his sound mirrors, scientist Robert Watson-Watt had been developing radar. His work had been prompted by the air ministry who wanted to know about the chances of developing a death ray which could take out enemy aircraft. Instead of knocking aircraft out of the sky, Watson-Watt decided to see if radio waves could be used to find out where they were so that fighters could intercept them. His system, which built on the theories of others, involved sending out beams of radio waves, then monitoring when they bounced off approaching objects. At first his system detected aircraft at 8 miles, then 40 miles, and eventually 200 miles. Almost overnight the sound mirrors had become redundant, and in 1939 the Acoustical Research Station was closed. The authorities even ordered that Tucker's remaining concrete mirrors be blown up. Thankfully, for military historians, these orders were ignored.

During the Battle of Britain, in the summer of 1940, the string of radar stations, which had been hastily built on the back of Watson-Watt's work, helped thwart Hitler's planned invasion. Using radar, enemy raids could be quickly pinpointed and intercepted by fighter aircraft. It also helped that the head of the Luftwaffe, Hermann Goering, decided to switch his strategy away from attacking radar stations to targeting the nation's cities instead. While destructive for Britain, these raids allowed more successful attacks on German bomber squadrons. Sir William Sholto Douglas, who went on to lead Fighter Command during World War Two, said: 'I think we can say that the Battle of Britain might never have been won … if it were not for the radar chain.'

Tucker's work had not been entirely in vain, however. During military exercises involving the sound mirrors in the early 1930s he had helped develop an early warning network to relay information on enemy activity to a central command. This concept, which was turned into a highly effective network with radar, was one of the key reasons why radar became such a useful tool for Allied forces in the Second World War and, arguably, helped turn the tide of the conflict.

5

THE DIABOLICAL
DEATH RAY

In 1916 an American named Albert Bacon Pratt was granted a patent for a new, hair-raising invention – the helmet gun. The firearm was, said Pratt, 'adapted to be mounted on and fired from the head of the marksman'. The user would be able to shoot instantly by tugging on a wire attached to his mouth. Usefully, continued Pratt, the device could also double up as a cooking pan. The idea was never heard of again – perhaps because the recoil of the gun would surely have broken the neck of anyone foolish enough to use it. Almost as bizarre was another US patent, filed in 1862, for a combined plough and gun. The basic idea was that you could get on with tilling the earth safe in the knowledge that you were fully prepared if marauding soldiers suddenly attacked across your fields.

It seems that mankind is always aiming to invent ever more fiendish ways of waging war. One 'Holy Grail' of warfare, however, has been the attempt to develop a so-called death ray. Greek inventor Archimedes supposedly used a 'burning mirror' death ray against Roman ships during the Siege of Syracuse in 214–212 BC. If he did do so, it can't have worked very well because the Romans ultimately triumphed in that contest. And it wasn't until the late nineteenth century and early twentieth century that the idea of an all-powerful death ray really caught the imagination, not only of science fiction writers like H.G. Wells, but also some of the brightest inventors of the day.

One of those who led the scramble to develop a workable death ray was the Gloucestershire born inventor Harry Grindell Matthews. He is a now half-forgotten figure. Yet in the early 1920s, his tussles with the British military establishment over the device became a newspaper editor's dream – a story that just ran and ran. Matthews, born in 1880 and married twice, once to an opera star, was certainly an eccentric figure. But he was also very clever. Initially an electronics engineer, he gained fame after inventing a wireless communication machine called the aerophone – a device which Matthews said was able to send messages to planes from the ground. Matthews even proudly showed it off to the then chancellor of the exchequer and future prime minister David Lloyd George. Matthews did early work on talking movies too, recording an interview with Ernest Shackleton in 1921 just before the explorer set off on a doomed final trip to Antarctica. A tireless polymath, his luminaphone was a machine that used beams of light to make music.

Matthews also knew a thing or two about battlefields, having served in the Boer War. In fact it was his determination, after being injured himself on the battlefield, that mankind should avoid more wars. This fuelled his efforts to create a weapon that, using a powerful electrical beam, could disable aircraft engines in mid-flight. If mastered it would provide each nation with the ultimate deterrent.

During the First World War the British government offered £25,000 to anyone who could come up with machines to remotely control aircraft or zap attackers in the sky, from the ground. They were particularly worried about the new threat from the air heralded by German Zeppelins, which killed over 550 people during raids over the country's major cities between 1914 and 1918. Matthews seized his chance to enter the fray. He developed a wireless boat which impressed admiralty officials when he demonstrated it on a London lake in

1915. Despite paying out the £25,000 prize, however, the government never developed the craft any further.

Later, in 1923, Matthews popped up again with a new invention to tackle the airborne threat as military aircraft were developing fast. He suddenly announced his idea of a death ray to a select group of journalists. They reported that in a demonstration his creation had successfully stopped a motorbike engine at a distance of 50ft. Matthews announced: 'I am confident that if I have facilities for developing it I can stop aeroplanes in flight.' In February 1924 The Air Ministry asked him to exhibit his death ray to their experts. Matthews refused. This may have dated back to his annoyance at allegedly finding a military official tampering with his famed aerophone back at another demonstration in 1911. Crucially, though, he wasn't forthcoming in divulging the science behind his death ray. In the ensuing media frenzy there was talk of ionized air and short radio waves, but Matthews kept up the mystery on how his beam worked while soaking up the limelight.

After some arm twisting by his backers, Matthews was finally persuaded to reveal his death ray in the spring of 1924. It was not the dramatic spectacle the ministry scientists had been expecting. In the lab his beam did appear to switch on a light bulb and cut off a small motor, but government officials weren't impressed. Some said the science wasn't new and, with rumours of con tricks doing the rounds, the government went cold on the idea. Matthews decided enough was enough. He would sell his invention to the French. His team of financial backers pleaded with him not to, even racing to the airport to try and stop him.

Meanwhile the press were having a field day with the saga. The *Daily Express* screamed of 'Melodramatic Death Ray Episodes' on its front page. Questions were being asked in the House of Commons too – shouldn't the government be making more effort to keep the death ray British? The government was forced to back down. It offered Matthews £1,000 if he could show that his ray would stop a motorcycle petrol engine. Matthews, now in France, retorted that he was doing a deal with the French. Indeed his associate there, Eugene Royer, filed a patent for a death ray around this time. In addition, that summer Matthews made a film with Pathé News showing off the shiny death ray machine and how it could kill a rodent at the flick of a switch.

Yet, strangely, a finished weapon never materialised, either in France or back in Britain. Later the same year, and still famous, Matthews set off for a fundraising tour of the US. Again, however, he refused to explain or show how his death ray worked. Over the next few years, interest in his claim and the offers of cash gradually faded away. Then, on Christmas Eve 1930, Matthews was back in the headlines. Above the skies of London festive shoppers could make out the words 'Happy Christmas' projected onto the clouds. It was the work of Matthews' new project – the sky projector. This time his beam wouldn't be killing things: it would help sell them. It seemed like a great advertising gimmick. What is more, there couldn't be any question that it worked. Sadly for Matthews it proved a commercial flop and he was declared bankrupt in 1934. When Matthews died in 1941 he had become a virtual recluse, locked away in a remote compound in Wales and still beavering away on new schemes.

Many more exponents of death rays had followed Matthews. In 1924 an American, Edwin R. Scott, claimed that he had actually been the first with a 'lightning device' that could 'bring down planes at a distance'. In the early '30s the Spanish-born scientist Antonio Longoria reckoned his death ray could kill birds 4 miles away. There were many other outlandish claims. In the end, none of these colourful characters were able to convincingly show that a death ray could work.

Despite this, even the renowned scientist Nikola Tesla, whose earthquake machine is the subject of the next chapter, believed it could be done. In 1934 he theorised that a concentrated beam of tiny particles formed of mercury or tungsten could be accelerated by a high-voltage current and used to knock whole squadrons of planes out of the air, hundreds of miles away. This teleforce, as he called it, would 'afford absolute protection' to countries from any form of attack. But Tesla, already in his late 70s, was to die before he or anyone else could turn his concept into a reality. Just a few years later, nuclear weapons would overtake the mania around death rays and, once again, they were consigned to the realm of fiction and fantasy.

6

TESLA'S EARTHQUAKE MACHINE

Famous in the early part of the twentieth century, Nikola Tesla deserves to be afforded more honour today. Arguably the inventor of radio (despite Marconi pipping him to the patent) as well as the developer of the alternating current (AC) – which pitted him against electricity giant Thomas Edison – he was the underdog in many battles. However, no one beats Tesla's supremacy in earthquake machines.

His pocket-sized device, the reciprocating mechanical oscillator, worked on the principle that every object has a 'resonancy' – a point at which it will begin to shake or shatter objects around it – very much like glass does in the presence of a piercing opera singer. With his experiments on high frequencies breaking new ground, such a machine could serve very practical purposes, such as discovering new oil fields or sites of archaeological interest.

By 1897, and now a big name in the field of electromagnetic discovery, Tesla had already shown how electrical energy could be transmitted wirelessly – something that is today known as the Tesla effect. He'd partnered with American electric company owner George Westinghouse to promote alternating current, and was working in a central Manhattan laboratory on new projects. Amongst them, he wanted to produce a device that, by resonat-

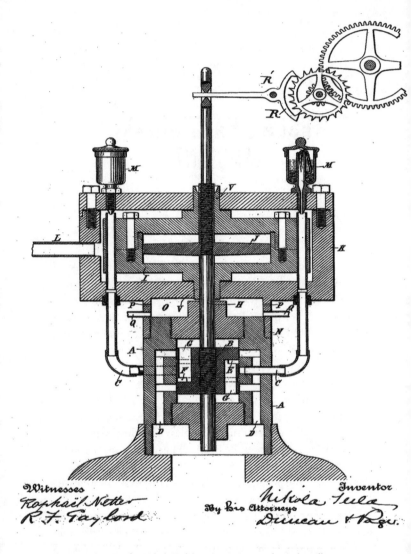

A patent drawing for Nikola Tesla's reciprocating mechanical oscillator or 'earthquake machine', which he claimed could be used to discover new oil fields.

ing in tune with the natural vibrations that all structures make, could shake buildings to their core. It wasn't long before he had a small machine that was up to the job of pulling things down.

Forever treading a fine line between sanity and madness, as Tesla turned on the machine in his lab one day he felt the walls shudder. Excitement got the better of him and he ratcheted up the power. 'Suddenly, all the heavy machinery in the place was flying around,' he claimed. 'I grabbed a hammer and broke the machine. Outside in the street, there was pandemonium.' New York's emergency services swung into action but could find no cause of the trouble. Tesla swore his assistants to secrecy, but the word on the street was that an earthquake had struck. And the cause: the Tesla oscillator, earthquake machine.

The story of Tesla's shaking lab has never been proven, but the inventor, eager to register a patent ahead of the rivals who had humiliated him in the past, was happy to embellish it, thus talking up the potential of his small, light, steam-powered device regardless. When suitably tuned, Tesla claimed, the oscillator could destroy anything. Resonance was already known to damage buildings and bridges. Now here was a gadget that pushed the laws of physics to their limits and beyond. 'Five pounds of air pressure … that is all the force I would need' to fell the Empire State Building, he said. 'With the same oscillator I could drop the Brooklyn Bridge in less than an hour.'

The technology was relatively simple. When steam is forced into the oscillator, the device vibrates until it reaches its natural resonance when shocks begin and the device then tries to shake itself apart. Attached to a building, each beat of the oscillator can be made to link to the structure's own natural vibrations. With each stroke, the force is magnified, and once the frequency of the resonance equals the time taken for vibrations to spread throughout the building, there's trouble. The lower the resonant frequency, the easier it is. Only a small force is needed to generate localised Armageddon, as the resonant effects are magnified and terrifying.

After his claim about a Manhattan earthquake, Tesla sought further empirical data. Strapping the alarm-clock-sized vibrating machine to a 2ft-long metal bar, he hit New York's streets in search of a construction site where half-erected steel frames could be turned into not-at-all erected mashes of mush. Near Wall Street such a building had reached ten-storeys high and, ignoring the workmen on the scaffolding, he quickly strapped an oscillator to one of the beams. Tesla's account says that the structure began to crack and that builders came down to ground level in panic. Police were called once more to investigate another earthquake, no doubt puzzled that central Manhattan may have suddenly developed a geographical fault line. Tesla slipped away with the machine in his pocket, delighted that he could have destroyed a steel building within minutes. Or at least, so he said.

While the patent and a model for Tesla's earthquake machine exist, there's actually little evidence to verify his stories. This seemed not to matter to him. His interest in making the earth move diminished as his public spat with Thomas Edison intensified during the first years of the twentieth century. Harbouring a grudge against Edison for the non-payment of a $50 bonus once promised if Tesla could improve upon Edison's inventions (he did), scientists and the press were by now debating the relative merits of Tesla's alternating current and Edison's competing direct current. By 1903, to Tesla's undoubted pleasure, AC had become the accepted standard in the 'war of the currents'.

Having become used to receiving royalty payments as electricity became more commonplace, coming second to Tesla had hit Edison in the pocket. To protect his invention and get back at Tesla, whose increasing oddness could be used to whip up concern about his alternating current, Edison decided he needed to convince the public about the inherent dangers of AC. For this purpose, he called upon the service of an elephant awaiting execution. This PR stunt proved quite a shock, not least to Topsy the elephant, and, when AC felled the beast, Tesla's alternating current suffered commercially.

Topsy wasn't the first animal to fall foul of Edison's need to prove his point. Smaller animals had already met their match (being 'Westinghoused' as Edison expressed it, a reference to Tesla's patron). Cats, dogs, horses, the odd cow; all were zapped by Edison in private. When news reached him that Luma Park Zoo at Cony Island was top heavy in elephants – Topsy in particular, treading keepers to death at the rate of one a year, was proving too much for her handlers – he was delighted. Topsy, the first and indeed only elephantine execution by electricity, may consider herself lucky. The zoo's owners had considered a hanging, but animal protectionists objected on humanitarian grounds. Instead, after being fed with carrots laced with cyanide, and in front of a crowd of 1,500, Edison zapped 6,000 volts of alternating current through the elephant from the legs up. Topsy died instantly – and Edison, satisfied that Tesla's current wasn't safe, got his press coverage. A recording of the event, *Executing an Elephant,* was released later that year (and is now saved for posterity on YouTube).

Finally zapped in the war of the currents, Tesla turned his considerable talents to other interests, including gas-powered airships (he dismissed the planes then flying as 'toys' that would never become commercially practical because of 'fatal defects'); a new kind of flying machine powered wirelessly from energy stations on the ground; and his own 'death beam'. As life rolled on, his state of mental health diminished along with his fame. He once said: 'One can be quite insane and think deeply.' For a man obsessed with the number three – he often walked around a block three times before entering a building, demanded three folded napkins at every meal and would only stay in a hotel room with a number divisible by three – this was deeply apt.

Rather ironically, the only major recognition Tesla received in his lifetime was the Edison Medal; the American electrical engineers' highest award. Marconi had won the Nobel Prize for physics in 1909 for the radio; and Tesla's spat with Edison meant that both were considered – and overlooked – for the prize in 1915. On his death, the FBI seized Tesla's paperwork, deeming many of his ideas too secret to release.

EDISON'S CONCRETE FURNITURE

Around the same time that he was executing elephants, Edison developed a plan to murder music. After the failure of his very first invention – the telegraphic vote-reading machine for US Congressmen, so slow that by the time it had rattled into action members had often changed their minds about which way to vote – Edison decided that he would henceforth only invent products for which there was a clear demand. Hundreds of ideas later, concrete furniture was on the way, and the piano was a leading item in the repertoire.

As his career became increasing successful, Edison had developed a narrow range of philanthropic ideals, including the desire to place a piano in the home of every American. To ensure affordability, expensive, resonant wood would be replaced by a piano framework of concrete. Perfectly tuneable, if somewhat immovable, it was hoped the instrument would extend America's musical talent – although Edison was far from the best judge of this. He'd been pretty much deaf since childhood; the result, he claimed, of being picked up by the ears onto a moving train by a helpful conductor – presumably, as Edison's brother suffered similar partial deafness, a family hazard. But Edison had a reason to develop concrete furniture that went beyond the philanthropic. As owner of the Portland

Cement Company, he produced more concrete than he knew what to do with. The work to which the factory was being put when he formed it in 1899 – iron-ore production – wasn't working out and its heavy industrial equipment was set for scrap.

Here Edison's entrepreneurial spirit came to the fore. Urbanisation was continuing apace in the US, yet the standard of housing for the working classes was generally poor. Edison, who had grown up in poverty, dreamed of ridding America of 'city slums' by building inexpensive concrete houses for workers. He aimed to construct a house that could sell for no more than $1,000. And not just concrete houses, but concrete

Inventor Thomas Edison with a model of his 'revolutionary' concrete house. He believed the innovation would provide plentiful low-cost housing and even designed concrete furniture to go inside the homes.

houses equipped with concrete furniture (at a maximum cost, he calculated, of $200 to kit out the home). An altruistic venture aimed at alleviating the lot of the poor, Edison envisaged workers relaxing in the evenings with a sing-song round their concrete piano. Concrete, he declared, 'will make it possible for the labouring man to put furniture in his home more artistic and more durable than is now to be found in the most palatial residence in Paris or along the Rhine'.

Within four years, Edison's factory was turning out cement on a massive scale: 3,000 barrels a day by 1905. Highly mechanised, with enormous kilns, it was soon one of the biggest cement factories in the world. Combined with his ready eye for an invention, Edison realised that as concrete was durable and flexible, at least before it set, if he reinforced it with a mesh of metal rods, he could make ever larger structures. Not only could he build cheap, good homes for workers, whole houses could also be moulded in one piece. This process, which cast an entire house from one mould, rather than the traditional way of constructing individual walls and bolting them together, was costly, but opened up another market as well – *expensive* concrete homes. Ugly but pricey, the wealthy could proudly invest in his properties too. But whether for impoverished workers or affluent executives, pouring concrete consistently and reliably isn't easy, particularly when the mould has to make individual rooms all in one go. The mixture of the aggregates and water has to be just right, and hydraulic force has to push the concrete vertically in parts. The delicate process didn't always run smoothly.

Nonetheless, commercial projects started to be commissioned: the world's first concrete highway, numerous buildings in New York City and the Yankee football stadium amongst them. The first of the philanthropic homes for workers were cast in Huxton Street in South Orange, New Jersey, in 1911. By 1917 just eleven concrete homes had been constructed, yet not one was sold, even at $1,200 (only slightly more than Edison's original aim); a bargain price for a home at the time. Writing in *Collier's Weekly* in the 1920s, architect Ernest Flagg said: 'Mr. Edison was not an architect. It was not cheapness that people wanted so much as relief from ugliness and Mr. Edison's early models entirely did not achieve that relief.' There were practical problems of living in a concrete home too: one can go through endless nails hanging pictures on a concrete wall and householders who wanted to adapt the accommodation discovered the impossibility of taking a wall out when the house is cast from a single mould.

With only some hundred concrete houses sold, but a factory to keep in action, Edison stepped up production of concrete furniture. At the 1911 annual convention of American Mechanical Engineers, he announced a plan for bedroom sets retailing at $5; baths, phonograph cabinets (record players) trimmed in white and gold, and a piano. Just like his elephant stunt, everything was done with an eye to press coverage. For one press wheeze, concrete phonograph cabinets were shipped with a sign saying: 'Please drop and abuse this package'. He told the *New York Times* that the phonograph cabinet provided better acoustic qualities than wood. The concrete piano housed a wooden soundboard – there was no getting away from that – and to Edison's partially deaf ear, the tone would be indistinguishable from the traditional instrument.

After all, unlike wood, which rots and splinters, concrete, he claimed, was perfect for furniture. Edison did concede, however, that concrete reinforced with steel rods would probably make for an uncomfortable sofa. So he set to work amending the constituency of his concrete, suffusing it with air to make 'foam concrete'; an oxymoron that fooled no one. The price – though less than half of that of equivalent furniture – was surely attractive. It would also be an investment; newlyweds would have furniture that would last longer than their marriage. It may be heavy, but Edison felt that was part of its attraction, and with the new foam concrete it would only be about 25 per cent heavier than wooden alternatives. It would look good too. With a bit of spit and polish and a special paint, concrete furniture could appear like any type of wood the purchaser required.

By the time he died at the age of 84, Edison had amassed 1,093 patents. He had invented or further developed products such as the incandescent light bulb; the phonograph; telegraph; a 'stencil-pen' (the forerunner of a tattoo drill); the sewing machine and the X-ray. Other intriguing ideas, such as a device for communicating with the dead, never got beyond the scope of his imagination. But Edison's concrete furniture, his concrete piano and his concrete houses are not amongst his most notable successes. In Huxton Street, some of his concrete houses still stand today: concrete, after all, being nothing if not enduring.

8

THE MISPLACED MAGINOT LINE

In the 1920s André Maginot was a worried man. A veteran of the First World War trenches, he'd seen his home town of Revigny-sur-Ornain in the region of Lorraine, near the border with Germany, bombarded. He had also been badly wounded in the leg at the bloody battle of Verdun on the Western Front. Maginot was determined that the 'rape' of France at the hands of its old foe would never happen again. He felt that the Treaty of Versailles – later blamed for leaving Germany in a bitter frame of mind that fuelled Hitler's rise to power – didn't go far enough, and that France was still vulnerable to attack from a Germany that simply couldn't be trusted. Maginot would dedicate much of the rest of his career to championing the idea of a massive line of fortifications along the border between the two countries. Fortunately for him, Maginot never lived to see it become one of history's most derided military concepts.

Maginot became a member of the French parliament when he returned to civilian life after the First World War and, by the late 1920s, was Minister of War. The idea of a defensive line had been around for a while. Ominously, Philippe Pétain, then Inspector General of the army, was one of its fans. As marshal, Pétain would later become infamous for leading the subjugated government which collaborated with the Germans following France's surrender in 1940. But a fortified line had its critics too. Most of the French communists and socialists were against it. Others, like Charles De Gaulle, would warn that it left France cowering behind its defences; he favoured a more mobile army. Maginot warned that French manpower in the 1930s would be no match for Germany's bigger population if conflict came, and that a defensive line would keep the peace by deterring an invasion. It would create jobs too and, if it had to be defended, could be done by smaller numbers of men rather than the massive armies which had been needed to repel the Germans in 1914–1918.

In 1926, funding was given for some experimental sections of line and in December 1929 Maginot outlined his plan for a full-scale project:

> We could hardly dream of building a kind of Great Wall of France, which would in any case be far too costly. Instead we have foreseen powerful but flexible means of organizing defence, based on the dual principle of taking full advantage of the terrain and establishing a continuous line of fire everywhere.

Maginot's arguments convinced enough French members of parliament and by 1930 he'd swung a vote to give 3 billion francs to roll out a massive line of fortifications along the country's eastern border.

When Maginot died on 7 January 1932, reportedly from complications caused by eating bad oysters, work was in full swing. During the decade of the Maginot Line's construction, four million truckloads of earth and concrete would be moved. The eventual fortifications, extending for hundreds of miles from Switzerland to Belgium, as well as along the Italian border, were impressive. Built of concrete and steel, the Line was not a continuous structure but involved a total of some 500 separate buildings: including a string of huge forts as well as artillery casemates, retracting machine gun turrets and tank traps, all close enough to be able to support each other. The main forts, known as *ouvrages*, each had a garrison of up to 1,000 men and often extended to several storeys underground. The fortifications housed everything from hospitals to living quarters connected by 100 miles of tunnels, even featuring underground railways. There were other fortified 'lines' like the German's Siegfried Line, later know as the 'Westwall'. But France's new defences were unmatched anywhere in the world and they proudly showed them off to visiting dignitaries.

In the end it was strategic errors that would ruin the Maginot Line's reputation, not the failings of these fortifications themselves. Belgium and France were allies, so the fortifications were light along their shared border leaving a large chunk of Northern France unprotected, should the Germans invade that way. The Belgians had their own defences and the forest of the Ardennes, which covered the area just to the north of the Maginot Line, was so impenetrable (so the thinking went) that the Germans surely wouldn't be able to pull off an invasion that way. If the Germans did come through Belgium then France's mobile forces could be concentrated in the area to see them off.

When Hitler finally turned his attention to the west in May 1940, his generals were, indeed, respectful enough not to try and attack the Maginot Line head on. Instead they came up with a strategic masterstroke, outflanking both the Line and the French and British armies distracted in northern Belgium. In a surprise strike they quickly captured Belgium's key Fort Eben Emael with an attack by paratroopers. Hundreds of Panzer tanks then swept through Ardennes forest, making a mockery of the myth that it was impassable. Within days the Germans were bearing down on Paris. Hitler's forces could now attack the Maginot Line to the south from the rear and, while many of its forts held out, by June France had surrendered. Tens of thousands of French soldiers were taken prisoner from the Maginot defences without having had the chance to fire a shot.

Like many great military blunders it's easy, with the benefit of hindsight, to criticise the Maginot Line. Debate rages among historians about whether it was ill-conceived and whether it worked. It certainly failed to save France. Historian Ian Ousby, for instance, has called it a 'dangerous distraction of time and money when it was built, and a pitiful irrelevance when the German invasion did come in 1940'.

Others have argued that the fortifications were so impressive that it gave rise to an atmosphere of complacency in France, a so-called 'Maginot mentality', so that it didn't plan properly for war or build a sufficient mechanised mobile force to back up the Line.

It's a common observation among military analysts that nations are 'always planning for the last war'. The Maginot Line certainly seems to fall into this category, devised by men who had lived through the largely static First World War in which 1.4 million of their countrymen had perished. They had not planned for the highly mobile nature of the Second World War with its blitzkriegs. Other arguments suggest that the Maginot Line wasn't the problem, it was the French army which hadn't been organised well enough, or that the air force hadn't been strong enough to support the fortifications. Whatever the truth, in one way, Maginot had constructed a 'Great Wall of France'. Despite what he had said, his Line shared something in common with the Great Wall of China. That great structure, for all its strength, didn't save the country from invasion either. In 1644 the Ming dynasty was overthrown when a rebel general merely opened a pair of its gates, allowing an invading army through to overrun the country.

9

THE GREAT 'PANJANDRUM'

The department of wheezers and dodgers had a problem. The Nazis' Atlantic Wall – a series of fortresses and batteries, bunkers and mines, running down Europe's west coast from Scandinavia to northern Spain – must be breached if Britain was to make headway in the Second World War. For the team charged with creating a weapon to breach it, this was a formidable task. Happily, the wheezers and dodgers (more formally known as the Directorate of Miscellaneous Weapons Development, DMWD) had an exemplary track record in creating contraptions of carnage. The Hajile, which dropped heavy equipment from the skies using rockets, and the Hedgehog, an anti-submarine bomb, were their inventions. And though the Atlantic Wall, which included concrete walls 10ft high and 7ft thick in places, would take some smashing at its strategic points, the DMWD had the talent, the willpower and the authority to meet the challenge.

Not least, they could call on the services of C.R. Finch-Noyes; full-time wing commander, part-time inventor, who had already developed 'hydroplane skimmers'. These short-range, high-capacity torpedoes were designed to obliterate Germany's Möhne and Eder dams by propelling themselves violently across water before reaching an exquisitely destructive end. While considerably less famous than the 'bouncing bomb' later developed by Barnes Wallis (and the subject of a renowned British war film), Finch-Noyes' missiles were the forerunner of Wallis' dambuster bomb. By developing such missiles, Finch-Noyes had learned how much more magnified explosions are when they occur underwater. He also believed in packing as big a punch as possible and so had increased the explosives in the warheads of torpedoes fourfold.

Puzzling over the Atlantic Wall challenge, he sketched an impression of the kind of weapon that would be required to hit thick, high concrete with sufficient force to breach an area large enough to get troops and equipment through. Most problematic of all was how to propel the bomb from landing craft at sea to the foot of the wall on the beach in the face of German gunners. His sketches turned into what became known as the 'Panjandrum', a reference to nonsense verse written almost 200 years earlier by actor Samuel Foote. Packed with explosives, Finch-Noyes' Panjandrum would nip out of the ocean from a landing craft and, once at the foot of the wall, blast all around to smithereens.

Masterminding the whole project was Sub-Lieutenant Nevil Shute Norway – who went on to write the novel *A Town Called Alice* under the name Nevil Shute – and he liked Finch-Noyes' design. An explosive wooden cart, with 10ft-tall wheels, powered by rockets and able to travel at 60 mph, hadn't been attempted before. This was the kind of impressive machine Shute needed: big enough to carry more than 1,000 kilograms of explosives and deliver its deadly load more or less in the right direction. Testing would have to be thorough and on terrain similar to the beaches of Normandy, where the Panjandrum was scheduled to be deployed first. Finch-Noyes' unit, the Combined Operations Experimental Establishment – or COXE as it was appropriately called in the circumstances – based at Appledore, Devon, knew of such a beach: Westward Ho! A west-country holiday heaven, with swathes of beautiful golden sand, high surf, sand dunes, a bit of shingle and some mud, it was everything that Normandy offered, and all on COXE's doorstep.

But secrecy, up to this point absolute, was compromised once the prototype Panjandrum, constructed in East London, hit the road en-route to the resort: the sight of an oversized cotton reel with two giant wheels and explosives in the middle raising eyebrows as the convoy trundled south-west. Like the Pied Piper, the Panjandrum attracted followers by the score, to the chagrin of the military. Ahead of testing, the DMWD implored onlookers to stay away for their own safety, to which the good people of Devon responded magnificently by turning up in droves to watch what many hoped to be the most extravagant firework display the town had ever seen.

Tests began modestly. Sand replaced explosives to start with, and low power rockets – just nineteen of them initially – were deployed for safety reasons. This was fortunate. On the first run, after careering erratically along the beach, the rockets fell off one wheel and the Panjandrum stuttered in every direction at the same time. On each subsequent test, control was tenuous to say the least. Nevil Shute Norway scratched his head and called for a pause for regrouping, which turned into a fundamental reassessment of the Panjandrum's design. What this weapon needed, he concluded, was another wheel.

Returning to the beach, the new and improved, three-wheeled, higher velocity Panjandrum carrying many more rockets, fared no better. The test team rolled the contraption towards the low-tide mark, cleared space around them, put their fingers in their ears, and fired. Nothing. They tried again and again. Still nothing. Even with the incoming tide lapping around their ankles and time running out, nothing. Finally, with the water all but engulfing the Panjandrum, testing was abandoned and the machine left deluged until the sea receded once more; the team looking on from the shoreline and wondering when the war would be over.

After three further weeks of modifications, the Panjandrum sported an improved steering system, seventy 20lb rockets, and agreement from top brass that accuracy wasn't altogether a pre-requisite when blowing up German walls. This time Nevil Shute Norway was confident. Firing the Panjandrum from a landing craft towards the beach, the test began well, until several rockets once again detached themselves and went haywire. Some narrowly missed the heads of startled spectators; others exploded underwater. Norway scratched his head again and decided that the third wheel was probably unnecessary after all.

A series of later test runs highlighted new problems, some more serious than others, but the team eventually declared themselves satisfied. One more test was required in front of their superiors, and in January 1944 military bigwigs left their Whitehall bunkers for the beach at Westward Ho! Focusing their binoculars on the landing craft, they may well have been giving themselves a small congratulatory pat on the back when the test got off to a credible start. The Panjandrum hurtled from the sea at wondrous speed, sparks flying from its rockets. But then old habits kicked in. As rockets dislodged, the Panjandrum spun, and shock replaced the confidence of those attending. Generals fled for cover. The official cameraman was almost mown down. And the Panjandrum, rockets flailing, wheels ablaze, disintegrated.

Shell shocked, the department of wheezers and dodgers called time on the Panjandrum and that looked like the end for it until, in a one-day-only revival in 2009, a Panjandrum was constructed once more. To mark the 65th anniversary of the D-Day landings, a slightly smaller and thankfully less explosive reconstruction of the calamitous launch took place at Westward Ho! With the Union flag fluttering proudly on the sands, a small crowd gathered to see if it could repeat the exciting experience of its wartime launch. As the rockets powered up in a Catherine Wheel display of smoke and flames, the situation looked promising. But the drum hardly hurtled off its ramp in a manner likely to frighten passing Germans, propelling itself just 50m before shuddering to a halt. It did, however, travel in the right direction, no rockets chased after cameramen and no one feared for their lives.

10

THE FIRST CHANNEL TUNNEL

Just 21 miles separate Britain from the coast of France at the narrowest point of the English Channel, or La Manche, as the French call it. Here, it is, on average, only around 120ft deep. So building a tunnel underneath this stretch of water was certainly not beyond the Victorians' capabilities. Much of the technological know-how about how to burrow beneath it and link the two nations did indeed exist a hundred years before today's rail tunnel was built. They nearly did it too. In 1880, using state of the art technology, boring began. Prime Minister William Gladstone and his wife even went down. As did the Archbishop of Canterbury. So what happened to this grand scheme and why was it another hundred years before digging began in earnest once more?

As early as 1802 Albert Mathieu-Favier, a mining professor, had put forward the plan of a tunnel during a short interlude in the wars between Britain and France. His idea would have seen horse-drawn carriages trundling through with oil lamps lighting the way. There would be a place to change horses half way through and enormous chimneys to the surface of the water giving ventilation. One cartoon print from the era has the Napoleonic armies triumphantly marching through a tunnel underneath to conquer England with hot air balloons flying across in support.

By the latter half of the century the technological expertise was in place to make the theory of a tunnel a practical reality. Soil surveys showed that the chalk under the Channel would be relatively easy to bore through. Other projects had already proved what could be achieved with modern machines. In 1843 Marc Isambard Brunel's tunnel, 1,300ft under the Thames, had been opened successfully. The 100-mile Suez Canal had opened to shipping in 1869, while an 8-mile railway tunnel had been built through the Alps in 1871 in what many engineers consider tougher geological circumstances. The aftermath of the Franco Prussian war of 1870 brought a tentative rapprochement between France and Britain, with Germany now seen as a common threat, creating a political environment conducive to a tunnel attempt. In 1873 *The Railway News* even ran an article saying that a Channel Tunnel was not only practicable but might pay for itself, too. But it would take private capital to give the plan the impetus it needed.

Step forward Sir Edward Watkin. Some years before his attempt to build a rival to France's Eiffel Tower (see Chapter 11) he was trying to build a railway link to the country too. Watkin, chairman of the South Eastern Railway, thought it a perfect opportunity to fulfil his dream of a railway that would eventually connect his Great Central line from Sheffield to London and right through to Paris. His Submarine Continental Railway Company aimed to tie up with a tunnel being built by a French group on the other side, led by Alexandre Lavalley, contractor for the Suez Canal. A trial tunnel was thus drilled near Dover in 1880. An inscription by one of the workers on the remains can still be seen today. It reads in rather appropriately faltering English: 'This Tunnel Was Begubugn William Sharp in 1880.' By 1881, the digging of a 7ft tunnel was well under way 100ft below the sea from Shakespeare Cliff, near Folkestone, with good progress being made on the French side, at Sangatte, near Calais.

The speed of the operation was helped by a new wonder – the Beaumont and English boring machine, which had a rotating cutting head to slice through the rock. Within five years, the pilot tunnel connecting the two countries would be finished, announced Watkin. By 1883 a total of 6,178ft had been excavated on the English side and 5,476ft on the French side. They were over a mile out from each coast. Watkin predicted 250 trains, powered by compressed air, could be going through the tunnel every day.

But military top dogs, most prominently a war hero called General Sir Garnet Wolseley, were getting twitchy. He even went so far as warning that legions of French troops could come over disguised as tourists then suddenly seize Dover. He warned a parliamentary commission into the affair: 'No matter what fortifications and defences were built, there would always be the peril of some continental army seizing the tunnel exit by surprise.' Watkin lobbied hard carrying out 'personally conducted tours and picnics'. But the generals had much of the press on their side, tapping into the public's ancient fear of invasion. The Channel had, many believed, safeguarded the country for hundreds of years. Even the *Railway News* was now dead against the idea.

The government subsequently caved under the pressure and in July 1883 the Board of Trade put a stop to the work. Watkin appealed to common sense, saying that if the government feared a French invasion, they would be able to flood the tunnel, blow it up or simply fill it with smoke and that, if necessary, elaborate fortifications could be built at the British end. His pleas fell on deaf ears. Watkin never quite gave up while he was still alive. Rumour had it that he had even sponsored Gladstone's trip to Paris in 1889 hoping that the tunnel project could be revived. It was to no avail. There was to be no Channel Tunnel out of England for another hundred years. For decades, military opposition and then economic constraints would sink several new ideas and suggestions for a Channel link.

As late as the 1980s, there were plans to forget building a tunnel and build a bridge instead. In fact the £3 billion suspension bridge, which would have been 220ft high, almost got the go ahead from the then prime minister Margaret

Thatcher. It would have seen motorists charged just £5.60 to cross between the two countries, taken ten years to build and involved the use of a massive 450,000 tons of steel. However, a Department of Transport memo from 1981 concluded that the bridge would be subject to too many safety and maintenance issues. The government finally opted for a subterranean link and the Channel Tunnel opened in 1994.

11

LONDON'S EIFFEL TOWER

When the Eiffel Tower was opened in Paris in 1889 it became the tallest man-made structure on the planet and made the city the envy of the world. Patriotic Briton Sir Edward Watkin thought Victorian London should respond with its own rival tower, vowing: 'Anything Paris can do, we can do bigger.'

Today, few of those who gather at that modern-day temple to lofty aspirations, Wembley Stadium, realise that they are sitting on the very site of Watkin's attempt to pull off his 1,150ft architectural riposte to the French. Yet in the early 1890s, just a few years after the 984ft Parisian landmark was inaugurated, it looked as if this audacious Liberal MP and railway magnate might get his way. For a magnificent steel edifice, planned to be over 150ft taller than its French counterpart, was beginning to rise majestically over the skyline.

Watkin was the quintessential nineteenth-century entrepreneur, who believed that modern engineering and technology could overcome almost any obstacle. Aged 72 when work began on the tower in 1891, he was already a successful man, even by the exacting standards of the industrial age. He had set up a newspaper; helped build railways in nations as far removed as Canada and India; had risen to become director of several railway companies, including Brunel's Great Western

Original 1890 design for
Sir Edward Watkin's tower,
which would have been
1,150ft tall.

Railway; and was a knighted member of parliament. While his
attempt to build the first Channel Tunnel had floundered, if
anyone could build a British Eiffel Tower, he seemed a good
man to have behind the project.

The site was to be Wembley, then a village of around 3,000
souls. As befits a true man of money, Watkin chose the location,
then in open countryside and some 8 miles from the centre of
the city, because it tied in nicely with his other projects. Watkin

Photograph showing the first, completed stage of Sir Edward Watkin's Wembley tower. The edifice was planned to be taller than the Eiffel Tower in Paris.

was already chairman of the nation's first underground line, the Metropolitan Railway, which had opened in the 1860s and in the 1880s was being extended – running north-west out of London. Opening a station at Wembley, and a pleasure park too, would attract people to use the new line as well as satisfy the nation's thirst for engineering and industrial prowess. He would also have been aware that the Eiffel Tower was bringing in a multitude of visitors and had repaid the cost of its construction in just one year.

Some 280 acres were purchased for the site of Watkin's Tower, sitting at 170ft above sea level. At first, in what was either a gesture of goodwill or breathtaking cheek, Watkin asked Gustave Eiffel himself to design his tower. Unsurprisingly that offer was declined, with Eiffel remarking that the French: 'would not think me so good a Frenchman as I hope I am'. Instead, in November 1889, a competition was launched for new designs for Watkin's tower, stirring much excitement. The designs poured in – sixty-eight in all – and were exhibited at the hall of Drapers' Company. One was modelled on the Tower of Pisa but without the famous lean; another was to be some 2,000ft high; one featured a spiral railway which would wind around the tower; and another was to house a 1/12th scale model of the Great Pyramid of Giza; one was even to be called: 'Muniment of Hieroglyphics Emblematical of British History During Queen Victoria's Reign Tower'. Alas, the judges decreed that none of the designs would work as they stood and Sir Benjamin Baker, designer of the Forth Bridge in Scotland, was drafted in to help make one of the designs, which sported eight huge, spider-like, iron legs, more practical.

Watkin formed the Metropolitan Tower Construction Company to oversee the work, which in total was projected to cost £200,000. The tower, due to be completed in 1893, would weigh in at a hefty 7,000 tons, with lifts transporting visitors to the top. Some 60,000 people were expected per day. Photographs from the time reveal a structure that looks very much like a half-built Eiffel tower. It was, however, to be thinner and with four levels. Visitors would not only get a bird's-eye-view of the capital and surrounding countryside, but the idea was that it would house everything from restaurants to theatres, with exhibitions and even a Turkish baths. At the top there would be an observatory and the park below was to have a boating lake and a spectacular waterfall.

By 1891 construction on what was now being called The Wembley Tower had begun. But subscriptions were poor and construction dragged. The press at the time had mixed feelings about Wembley Park. The *Pall Mall Gazette* opined: 'The chief attraction is the Great Tower now in course of construction.' And the *Daily Graphic*, of 14 April 1894, noted that 'over a hundred and fifty men are constantly employed in putting the tower together'. *The Times* talked breathlessly of how an express lift would take impatient visitors to the top in just two and a half minutes. But an article in the *Building News*, also published in 1894, called the tower an 'unfinished ugliness'.

Eventually the first level was completed and lifts installed. In 1896 the tower was opened to the public. It had already reached a height of 155ft and was dominating the surrounding landscape. In the same year a local brewer was using a picture of the tower on advertising to sell their wares. Then, things began to go drastically wrong. It emerged that the foundations of the tower had begun to move. Poor surveying had failed to identify that the marshy ground could not support the weight of the tower with just four legs. Nor had Watkin realised that, unlike the Eiffel Tower, his tourist attraction was at that time too far from the city centre to bring sufficient visitors to pay for its construction, even though his own new railway extension ran right past it. Apart from the station and the parkland, many of the much-vaunted facilities had failed to materialise, meaning that in its unfinished state the tower wasn't much of a draw.

Money began to run out, and with only £87,000 raised by 1899 the company went bust and work stopped. There the tower remained, gently rusting, a sorry reminder of the failed venture. In the coming years it was used for army signalling and the odd firework display and dubbed the London Stump or Watkin's Folly. Finally, on 7 September 1907, it was blown up. The 2,700 tons of scrap metal that had gone into its construction were exported to Italy.

The site's famous days weren't over though. In 1923 Wembley Stadium – with its iconic twin towers - was built on the site. And remains of Watkin's structure were unearthed when the new Wembley Stadium, which opened in 2007, was being built. All that is left of Watkin for posterity is a Watkin Road, a few streets from the current stadium, lined with industrial premises in stark contrast to the local 'sylvan scenery' of the area described by *The Sunday Times* in the 1890s.

Watkin himself died in 1901, shortly after it became obvious that his dream had crumbled. Two years before his death, however, he was present at the opening of another more successful venture – Marylebone Station, the terminus for the Great Central's London Extension, one of the last mainline railways to be built in Britain. Despite his north-London folly, his railways – and Wembley Park station – are still in use by millions. To this day hoards of football fans use his line to flock to the spot carrying their own hopes of glory.

NELSON'S PYRAMID

Trafalgar Square, at the very centre of London, contains what is almost certainly the most famous statue in the country: a pillar, 170ft high with a 17ft-tall Horatio Nelson on top. But the square might have looked very different if the judges of a competition to design a monument to one of the country's greatest seafaring heroes had matched his bravery.

The options in front of the Nelson Memorial Committee varied from the absurdly exotic, such as a replica of the Roman Coliseum, to the ridiculously trivial in the form of a pair of marble dolphins. Often Nelson featured in numerous shapes and sizes; occasionally he was replaced with an attractive alternative, perhaps a seafarer stripped to a muscley waist, maybe three nymph-like mermaids. But long before the competition, a plan for the site was put forward by an enterprising, energetic and eccentric Irish knight of the realm, Colonel Sir Frederick William Trench MP. His eye-catching idea: a British pyramid.

At the time that Nelson met his end at the battle of Trafalgar in 1805, his now eponymous square was known as the King's Mews, housing only unpleasant dwellings for the poor and fine horses for the king. And although the recently deceased Nelson was already written into legend, thought didn't turn to a permanent memorial for a generation. In the meantime,

Colonel Trench decided that the thrashing of the French, not the beatification of Nelson, was something that should be monumentally marked in the capital.

In 1815, Trench began his campaign with the same interfering style which had been making him a nuisance throughout London for much of his life. Dismissing plans for a new royal palace in Whitehall as 'narrow tradesman-like views', he upset the aristocracy, architects and the workers in one deft sentence. Despite this, Trench was, he claimed: 'an advocate for the splendour and magnificence of the crown', and in this respect he was determined to spend the nation's crowns on a dramatic structure, London's largest, in what was to become Trafalgar Square.

The pyramid which might have been erected where Nelson's column stands today. Designed by Philip and Matthew Cotes Wyatt, it would have cost £1 million.

Trench thought the pyramid would be the nation's most celebrated building, exceeding the height, the footprint and the glory of St Paul's Cathedral. With each of its twenty-two stepped piers representing a year of the recent wars with France, the structure would remind a grateful nation how Bonaparte's fleet, superior in numbers but lesser in talent, commitment and honour, had been defeated in the Battle of the Nile. After Napoleon set his sights on building an empire along British lines, the Middle East campaign became a turning point in the Anglo-French wars, with the French leader taking Malta on his way to seize Egypt. Napoleon's troops took Cairo easily, but then Nelson's stunning attack at Abu Qir Bay, off the port of Alexandria on 1 August 1798, overwhelmed most of the thirteen French ships. Just four escaped. Balancing Nelson's regret that he had missed a full house, British cannons had happily sliced off the right arm, then the left arm and then one leg off one French captain, Aristide Auber Dupetit-Thousars, who nonetheless continued commanding his ship after being placed in a tub of bran. This, to the one-armed, half-blind Nelson, was sweet justice.

Colonel Trench believed this was the Napoleonic battle to commemorate. Trafalgar may have been the decisive battle of Anglo-French wars, but it proved fatal for Nelson. Egypt, by contrast, was a more exotic event in a greatly romantic setting, in which Nelson only suffered a head injury and was later able to report his 'irresistible' pleasure at the discipline of his officers and men. All in all, thought Trench, this was a battle worth spending £1 million to remember. The pyramid's erection was 'an expense not burthensome to the nation' he told incredulous sponsors while asking for the money to cover the cost.

To prove the seriousness of his intentions, Trench commissioned a model from architects Philip and Matthew Cotes Wyatt, sons of James Wyatt the Fonthill Abbey designer (we will meet in Chapter 14). Needing big names to support the plan, he asked the former commander-in-chief of the army, the Duke of York, second son of mad King George III, to provide royal patronage and allow his Pall Mall residence to be opened for an exhibit. The duke, not known for sound judgment or strong intellect (he was an incorrigible drunk, womaniser and gambler in the finest traditions of the Hanoverian monarchy), agreed. To the great loss of a nation in need of a pyramid, the exhibition failed to attract the support Trench expected. A busybody, eccentric MP, joining forces with a stupid, drunken duke (it was he who is often credited with marching 10,000 men to the top of the hill and marching them down again) and asking for a cool £1 million for a monument that would dwarf St Paul's, impressed few. Nelson's pyramid got no further than the model in the Duke of York's home.

But that didn't stop changes to King's Mews. By 1830, horses and houses had been cleared away to leave an open space, the name changing to Trafalgar Square five years later. The whole area around Whitehall by this time resembled nothing more than a building site: the nearby Houses of Parliament burning down in 1834 just as King's Mews lay fallow. Construction of a lasting Nelson testimonial remained years away, but at last things were moving. A committee of the great and the good began to organise a competition and to raise funds for a statue, dolphin, Coliseum or whatever it was going to be, although certainly not a pyramid.

Respondents were invited to think freely, but when it came, the choice of a dull but functional and unquestionably tall column outraged almost as many people as had gasped at Trench's pyramid. However, Nelson's column it was to be, despite strong protests that dogged it throughout construction between 1840 and 1843. A disproportionately small Nelson, a tenth the size of the column he would stand on, was a folly that the public would not take to its heart, some said. No one would be able to see the great hero, said others, with not a little justification. It would spoil the view of Whitehall and parliament from the north. Even a parliamentary select committee joined the column bashing, but its hands were tied: 'We do not see how Government can avoid interfering to prevent the miserable result of which they are so clearly forewarned … Alas for the prospects of the Art in England.'

Today, London has three notable monuments to this period in British history. Nelson's column, but more broadly Trafalgar Square has, after all, won its place it the nation's affections. And the grand old Duke of York, the man who housed the model of Nelson's pyramid, has a 123ft-tall column of his own, paid for by a compulsory tax on British soldiers, each of whom had to give up a day's wages to fund it when he died. Frederick Augustus Hanover, Duke of York, is on the Duke of York Column in Waterloo Place, a quarter of a mile away from Nelson. And on the Embankment rests Cleopatra's Needle, a gift from the Egyptians to the British in 1819 that honours the very same battle that Trench wanted to remember: Nelson's Battle of the Nile.

13

WREN'S MISSING MARVELS

In September 1666, as diarist Samuel Pepys wept on the south bank of the Thames watching London burn on the opposite bank, an opportunist astronomer hatched an extravagant plan. From the ashes of the city – a city that, up until that point, had been a higgledy-piggledy mess of urban depravity and privilege in equal measure – he would build a new capital; the most majestic capital city in Europe. Christopher Wren's qualifications for such a task may have been lacking – at the time, his most significant brush with architecture was upsetting and almost coming to a fist fight with the designer improving St Paul's Cathedral – but he made up for his deficiencies with enthusiasm, intelligence and force of personality.

Recently returned from an extended tour of Paris, a catastrophe of biblical proportions presented Wren with a never-to-be-repeated opportunity: the chance to create a new city on a blank, and rather charred, canvas. To this point, Wren's career, though distinguished, was studious but not practical. Still in his 30s, as professor of astronomy at Oxford University, he had reached the pinnacle of English academia. He'd dabbled in poetry; his best attempt being twenty lines of florid verse about the resuscitation of a hanged woman. And he had produced a series of relatively unimpressive inventions: a device for writing in the dark; some graphite for lubricating timepieces; and a demonstration model of how a horse's eye worked. His experiments with animals – including removing the spleen of his spaniel to see how it would get on without one – had become tiresome. It was time for a career change.

So Wren, and London, had a lot to thank Thomas Farryner for. At about 1 a.m. on Sunday 2 September, Farryner, a baker who supplied the navy with ships' biscuit, noticed a smell emanating from elsewhere in his home in Pudding Lane. It was the Great Fire of London, and it was burning in the room next door. Waking his family, Farryner managed to scramble to safety, leaving only a housemaid, too frightened to flee, to the mercy of the flames and to become the fire's first victim.

London's fire risk had long been evident. Tinderbox wooden homes, lit by candles, heated by open fires; buildings packed tightly together; with fire-fighting reliant on buckets of sand, milk and, less convincingly, urine. After two days of chaotic attempts to extinguish the blaze, the situation looked worse, not better. On Tuesday, St Paul's caught light, causing Pepys, who had been following the fire's progress in his diaries, to feel the situation serious enough to bury his papers, his wine and a rather splendid parmesan cheese he had been working through. By Wednesday, with the blaze finally out, about 200,000 people were homeless. More than 13,200 houses, a hospital, the Royal Exchange, three city gates and, to the delight of inmates, Newgate Prison had been destroyed: 80 per cent of the city reduced to embers. Although only nine deaths were directly attributed to the fire, the true loss is believed to be many more as people fell victim to crime, hunger and poverty.

As debate about the cause raged – 666 being the number of the beast, in 1666 sin was the most obvious culprit – Wren progressed with his plans. Within days, he was in front of the king, Charles II, presenting a breathtaking design for a new London based on classical Paris and Rome. He had been wide-eyed with awe at the architecture in the French capital. In contrast to London's chaotic and still largely medieval streets, Paris, though smaller, was blessed with expansive boulevards, pleasant parks and elegant public spaces. Through his friend and competitor John Evelyn, with whom Wren almost certainly shared ideas before or even during the fire, he knew too of 1580s Rome under Pope Sixtus V, with its Palladian buildings, fine avenues and wondrous piazzas.

Wren's new London borrowed from both cities. The main boulevard was to run from Fleet Street to Aldgate, with a circus at the centre. A new Royal Exchange, rebuilt on its original site, would dominate a piazza where ten streets intersected. All the city's main commercial buildings – the Mint, Post Office, goldsmiths, the Excise Office, the banks – formed a commercial district. A second large avenue, with two piazzas along the way, started at the Tower and converged at St Paul's. The cathedral itself would be smaller than its predecessor and the number of churches elsewhere cut from eighty-six to fifty-one. Even the streets proclaimed sense and order, with a grid system off the boulevards banishing forever the cramped, unhygienic conditions of London before the fire. In his biography *Parentalia,* Wren's son, also Christopher, claimed the new streets had 'three Magnitudes' – thoroughfares at least 90ft wide, secondary streets 60ft, and lanes 30ft. Wren's scale drawing doesn't back up that claim, but the roads were certainly of a scale unseen in England before. A new era was coming.

Wren's and Evelyn's plans, though developed separately, shared much in common. Wren's was the finer, but it contained two fatal flaws. Firstly, it was based on inaccurate plans of the area, an elementary mistake for someone who was to go on to be the most eminent architect in British history. Secondly, and much more difficult to solve, it would take a very long time to build. Wishing to keep the people on side, the king wanted normality returned quickly. Wren's scheme, if it was presented with any seriousness at all – and that, from an astronomer, was in doubt – would involve undue delay. Moreover, Wren hadn't taken account of existing ownership rights. People would have to be compensated for having their land sequestrated. And the city was skint.

Just days after both Wren and Evelyn presented their plans, Charles vetoed them both as too complicated and costly. Most of London's income came from property, and most of that property was in ruins. The only way of getting the city back on its feet was to allow people to rebuild their houses and return to commerce, not faff about sipping coffee on a boulevard like they did in Paris.

Other architects rushed to Whitehall with schemes of their own, including Wren's friend Robert Hooke, and one Captain Valentine Knight, whose over-enthusiastic suggestion that the king might be on to a nice little earner from the rebuilding resulted in a spell in prison. Not one of the plans was implemented. From the safe distance of half a century, Wren's son attacked the king for not possessing sufficient foresight to understand the transformative benefits. Wren's own transformation from astronomer to architect was nevertheless under way. Along with Evelyn, Hooke and three other eminent men, the king appointed Wren to a commission which was to be funded, in the most British of ironies, with a new tax on coal – the rationale being that most people used it and it was hard to smuggle without the offender ending up sooty.

The six commissioners argued that London could still have splendid new boulevards and buildings, despite the cost and delay. But while the king suggested a middle way – a new quay and wider streets – when parliament debated the issue they quickly got bored and decided their time would be better spent working out how to escalate a conflict with Holland. (The Dutch were rumoured to be arsonists who had started the Great Fire in the first place in revenge for the English burning one of their harbour towns just weeks before.) All building plans were scaled back, the work was divided out and eventually Wren – later Sir Christopher – went on to design all fifty-one new parish churches in the city, together with his new landmark, St Paul's Cathedral. This, at £265,000, cost just over a third of the money raised by the coal tax. A tribute to the Great Fire designed by Wren and Hooke was erected near the place where the Great Fire ignited. But even this was not the monument that Wren first suggested. He originally wanted a phoenix on top of a column, but he decided that people wouldn't be able to identify the bird 200ft high above them. A statue of Charles in Roman costume didn't pass muster with the king who thought it would be too expensive, and why didn't Wren think of something along the lines of a large ball of metal that people could see at a distance? So the monument today is a Doric column with a flaming gilded urn.

The story does have a partially happy ending, however, for Wren's new London does exist in part in one capital city. Working from his original engraving, in 1799, US President Thomas Jefferson laid out what they then called Federal City. Today, Wren's London grid forms the basis of a part of Washington DC.

14

THE TUMBLING ABBEY HABIT

He was 'England's wealthiest son', a man of such means he could build a cathedral to live in if he wanted and get the greatest architect of his generation to design it. This combination of unlimited money and unrivalled talent should have resulted in a secular abbey so well constructed that it would survive the centuries; so beautiful that it would compete for the nation's affections with Salisbury or Winchester or York. Yet William Beckford's abbey at Fonthill, Wiltshire, which spanned 270ft from east to west and 312ft from north to south, with a central tower reaching almost 400ft to its turrets, stood for just a generation.

The whole enterprise should be a lesson to all homeowners who, thinking they know better than their professionals, over-rule them on crucial issues, change their minds after the plans are agreed, and who bribe builders with liquor. For Beckford did all this and more. Spoilt, eccentric, yet not unkind or unintelligent – he had an important art collection and was the author of bestselling gothic novel *Vathek* – his intention was to build an impenetrable structure where he could shut himself away from a hostile world and live out his days in peace.

Beckford had reason to be reclusive. His aristocratic ancestry meant he moved, when the muse very occasionally took him, in important circles. Blood of Edward I, Edward III, the House of Lancaster and the House of Stuart ran through his veins and the family was extremely well connected. When young William showed an interest in music, his father summonsed Mozart to England to provide lessons. 'He was eight years old and I was six,' Beckford recalled. 'It was rather ludicrous one child being pupil of another.' But the chemistry was good; so good that Mozart used bars by Beckford in the *Marriage of Figaro* – at least according to Beckford. But just four years after Mozart's lessons, Beckford's father was dead, leaving William an estate of £1 million. It was a sum – the equivalent of about £200 million in 2011 – that in 1769 could go a very long way indeed.

So precocious, talented, influential, landed and now just about the wealthiest person in the country, William Beckford had everything going for him. Then an indiscretion at the age of 23 (he had fallen in love with the 11-year-old heir to the Earl of Devon) led to him being ostracised from society and fleeing the country. This landmark event in his life arguably formed the origins of Fonthill Abbey. Given his high pro-

A view of Fonthill Abbey, William Beckford's enormous gothic home which stood for just thirty years.

file, and despite the allegations never being proven in court, Britain's newspapers, ever eager for a scurrilous story, went to town. Beckford fled to the continent with his adoring wife Lady Margaret Gordon, where she subsequently died in childbirth. This tragedy, combined with the earlier scandal, turned an already eccentric oddball into a recluse. With his wife dead, time on his hands and money in his pocket, Beckford decided to travel through Europe, taking along his doctor, baker, cook, valet, three footmen and twenty-four musicians. He wasn't an easy guest or boss. Hotel managers were required to redecorate before he arrived, and he expected exceptional service, however impractical: even if it meant shipping a flock of sheep from England to Portugal to improve the view from his window.

After thirteen years of this nomadic existence, Beckford decided sufficient time had elapsed for him to return safely to Britain. Back at Fonthill, he tore down his boyhood home, itself a Palladian mansion erected by his father only relatively recently, and constructed a very high and extraordinarily long wall behind which plans for the abbey began. He was, according to the *Country Literary Chronicle*, 'determined to produce an edifice uncommon in design, and adorn it with splendour'. With privacy uppermost in his mind, Beckford was anxious that Fonthill Abbey should be constructed in secrecy, although the sudden appearance of a 6 mile wall – 'the barrier' as it became known – had generated considerable local interest. A small village sprang up to house workers and the area buzzed with excitement. Villagers knew something big was happening and the secrecy added to the intrigue.

Former surveyor to Westminster Abbey James Wyatt, King George III's favourite architect, was delighted to win the commission for a project of such magnitude and, not to put too fine a point on it, with such an obscene budget. The decision to hire Wyatt, an architect who was much in demand, in itself had consequences for the subsequent fall of Fonthill. Wyatt was unable to give all his ongoing projects the attention they deserved and, in the opinion of Prime Minister Lord Liverpool, was an 'incurable absentee; certainly one of the worst public servants I recollect'. More tellingly, Wyatt had earned the nickname 'The Destroyer' after less than exemplary work at Durham, Hereford, Lichfield and Salisbury cathedrals, a slander that Beckford was happy to overlook, for without doubt, Wyatt's plans for Fonthill were exemplary. Towers, spires, cavernous rooms, enormous windows capturing splendid vistas, Fonthill was perfect – while it stood. It was the workmanship, not the design, which turned out to be shoddy. When Beckford tried to cut corners, Wyatt, for whatever reason, but presumably both absent and exasperated with a contrary client, let him have his way.

The more the builders worked, the more Beckford rushed them. Soon 500 men were employed on the construction, and when the nights began to draw in, they were ordered to work round the clock, by torch and lamplight. He almost doubled numbers by enticing builders working on Windsor Castle, one of Wyatt's projects for George III, with the promise of beer.

Even then, construction, thought Beckford, was stalling. That Europe's grandest abbeys had taken centuries to build was of no consequence. The nineteenth century was upon them and Lord Nelson was coming for dinner. Importing granite or marble or other fine material would waste time. Compocement and lime mortar that could be rendered with sand to look like stone was ordered instead, with the considerable downside that this substance wouldn't hang together sufficiently to hold parts of the abbey upright. Boozy builders, an impossible timescale and inadequate material rarely produce robust results. It wasn't long before parts of the building began to crumble. The first tower, having reached 300ft in height, collapsed initially during a gale. When Beckford ordered it to be rebuilt using the same material, it collapsed again. It was only on the third rebuilding that Beckford agreed to switch to stone.

But once finished, Fonthill Abbey was everything Beckford had dreamed of: magnificent and private. With a much-reduced retinue of servants, including a dwarf whose job description included standing by the 38ft high doors to make them look even taller, Beckford continued to write and collect art. A select number of old friends were invited to inspect the building. 'Who but a man of extraordinary genius would have thought of rearing in the desert such a structure as this, or creating such an oasis?' wrote Henry Venn Lansdown, not realising that by mounting the circular staircase around the towers to admire the view towards the Bristol Channel, he was taking his life in his hands. Nelson eventually visited with Lady Hamilton in 1800, and found himself in the unusual predicament of being publicly criticised for enjoying the hospitality of a man who had once been accused of lewd conduct.

Then another personal disaster struck Beckford. Poverty – at least in relative terms. The collapse of the sugar market had hit his Jamaican plantations, where 1,200 slaves worked the land. Suddenly he needed cash. Fonthill went on the market and the eventual buyer, a gunpowder dealer, was happy with his acquisition, until masonry began to tumble around him. In all, Fonthill Abbey's towers collapsed at least six times. The Great Octagon took its final fall on 21 December 1825 while a subsequent owner, John Farquhar, was elsewhere in the abbey. It was never to be rebuilt. By the 1840s, much of Fonthill was gone forever, the building left in 'desolate ruin'. After the death of a further owner in 1858, most of its remains were demolished. Today, part of the north wing which escaped the fall of the tower, and which has its own 76ft-tall turret intact, survives. But Fonthill Abbey, in all its magnificence, had stood for just thirty years. It was one of the shortest-surviving full-sized secular cathedrals in history. William Beckford himself retired to Bath where he bought two adjoining properties – connecting them with a new bridge and building a tower on the hill at the end of the road, which is now a museum in his honour. He is buried in the graveyard there, in a pink sarcophagus next to his dog.

15

WHY LUTYENS'
CATHEDRAL VANISHED

The Second World War saw the destruction of many
magnificent buildings. In Britain, the loss of Coventry's four-
teenth-century cathedral to bombing, is still mourned. But it
wasn't the only great cathedral to fall foul of the conflict. In
fact, thanks to the Nazis and the resulting age of austerity, what
might have been the country's most impressive place of worship was never built at all.

Liverpool's lost Roman Catholic cathedral would have
boasted the world's biggest church dome – a breathtaking edifice which would surely have been one of the architectural
wonders of modern times. Today the only part of the structure which was actually built, the crypt, lies beneath the more
humble, modernist, Liverpool Metropolitan Cathedral. Another
beguiling glimpse of what could have been there is still possible, however, thanks to the careful restoration of an amazing
12ft-scale wooden model of the original concept, now on view
at the new Museum of Liverpool. It reveals, in intricate detail,
just how ambitious and awe inspiring it was designed to be.
The cathedral's spectacular 510ft-high dome, measuring 168ft
in diameter, would have been bigger than that of St Peter's in
Rome. At more than 600ft long and 400ft wide, the building
would have taken up a massive 6 acres, twice the area as St

A model of Lutyens' Liverpool cathedral that never was, restored at National Conservation Centre, Liverpool, and now on display at the new Museum of Liverpool.

Paul's in London, radically transforming the city's skyline. Constructed from brick and silver–grey granite, it would have been entered through a huge, soaring arch.

The person behind this daring blueprint was one of Britain's greatest architects – Sir Edwin Lutyens. Sir Edwin first made his name designing Arts and Crafts-style country houses. He went on to create the Viceroy's Residence in New Delhi and many of the finest First World War memorials, including the Cenotaph in London's Whitehall and the Memorial to the Missing in Thiepval, France. At the height of his reputation in 1929 Lutyens, an Anglican, received what was to be his grandest commission – a Roman Catholic cathedral for Liverpool, set to be the crowning glory of his career. The city already had a new Anglican cathedral designed by Giles Gilbert Scott, himself a Roman Catholic, which was well on its way to being completed. But Liverpool's large Roman Catholic population, which had surged to 250,000 in the decades following the Irish potato famine in the mid-nineteenth century, was crammed into tiny churches, without a defining focal point of worship.

The first attempt to build a Catholic cathedral in the city had failed back in Victorian times, leaving only a Lady Chapel, later demolished. In 1928 the new archbishop, Richard Downey, was determined that Liverpool should get its second cathedral. He sought out Lutyens, the most famous British architect of his day and, according to Lutyens' autobiography, the pair arranged the deal to build it over cocktails! As the project was launched in 1929 Downey announced:

> Hitherto all cathedrals have been dedicated to saints. I hope this one will be dedicated to Christ himself with a great figure surmounted on the cathedral visible for many a mile out at sea.
>
> We do not want something Gothic … The time has gone by when the Church should be content with a weak imitation of medieval architecture. Our own age is worthy of interpretation right now and there could be no finer place than a great seaport like Liverpool.

Brownlow Hill, the site identified for the project, had once housed a huge Victorian workhouse. Now, bought for £10,000, it would be the setting for a place of sanctuary.

The scale of Lutyens' grand design was remarkable. Architectural historian Sir John Summerson summed it up as: 'the supreme attempt to embrace Rome, Byzantium, the Romanesque and the Renaissance in one triumphal and triumphant synthesis'. Estimated to cost £3 million, it would be 180ft higher than the neighbouring Anglican cathedral. Along with its dome, the complex structure was to be topped with belfry towers, spires and scores of statues. Inside, granite-lined nave and aisles would feature a series of barrel vaults running at right angles, and there was to be a total of fifty-three altars. Inside the West Porch a huge space called the 'narthex' would offer a place for the 'cold and destitute' to shelter. Even the church's organ was to be the world's largest.

Like the architects of most great medieval cathedrals, Sir Edwin knew that he probably wouldn't see the cathedral completed, believing it would take decades to finish. Downey however was in a hurry – and talked of a 'cathedral in our time'. Lutyens unveiled drawings of the cathedral to the public at the Royal Academy in 1932 and his grand £5,000 model of the cathedral was begun. Shortly afterwards Pope Pius XI gave the plans his blessing, and on 5 June 1933 (the year Adolf Hitler was elected in Germany) the first foundation stone was laid at an open-air mass. Work began on the foundations and by the outbreak of war in 1939, an incredible 70,000 tons of earth has been excavated and 40,000 cubic feet of granite had been laid. The crypt was almost finished. Only half of the growing £1 million fund, mostly raised from parishioners, had so far been used.

In 1941, however, construction stopped. Instead of a gracious cathedral, barrage balloons loomed over Merseyside; the area suffered seventy-nine bombing raids during the war with a death toll of 4,000 people. The crypt was used as an air-raid shelter. By the time the war was over, Lutyens had died from cancer.

When the project was finally revived in the early 1950s, costs had risen to a massive £27 million. Archbishop Downey died in 1953 and the new archbishop, Dr William Godfrey, aimed to scale back the original plans. In another twist to the story Adrian Gilbert Scott, brother of the Anglican cathedral's architect, was hired to come up with a design that kept the iconic dome but was much smaller and would only cost £4 million. But the resulting design was still deemed too pricey. In 1960 a competition was announced for an entirely new cathedral, which would cost a frugal £1 million. Lutyens' vision was to be finally consigned to history, and in 1967, Sir Frederick Gibberd's circular, concrete cathedral, since dubbed Paddy's Wigwam, was finished. The then archbishop admitted: 'You can loathe it, or you can love it – but you can't ignore it.' In the years that have followed the structure has been beset by leaks and corrosion.

The stunning model of Lutyens' cathedral is a stirring remnant of the church many would have preferred to see. For Sir Edwin's son, Robert Lutyens, the very fact that his father's lost Liverpool landmark remains only a dream is part of its appeal. He believed it had been saved from 'prejudiced denigration' in its unfinished glory. In a 1969 interview he explained: 'It is there, yet it is nowhere. It is architecture asserted once and for ever – the very greatest building that was never built!'

NEW YORK'S
DOOMED DOME

With inventions that crossed into the realms of science fiction, including plans for a Spaceship Earth, 4D towers and a fish-shaped car that was steered by a rudder, Richard Buckminster Fuller was one of the twentieth century's most intriguing inventors and a man with an attuned social conscience.

Although many of his ideas never saw the light of day, the 4D concept – referring to how the building would, in addition to its physical three-dimensional properties, be used over time – is a fundamental part of urban design today. This fourth dimension can improve energy efficiency and environmental impact, and, at its best, alleviate poverty and inequality. More improbably, his original 1920s' prefabricated tower blocks were designed to be flown into position across the US by airship. That didn't happen, but the lessons learned during writing the patent application for them would eventually be used in the development of 'geodesic domes' – large golf ball-like structures that are today widely used for science and entertainment venues. One dome in particular was set for big things: it would cover a huge part of central Manhattan, be energy efficient and have its own climate. Two miles in diameter and one mile high at its centre, the domed city would cover fifty of Manhattan's central blocks and benefit from its

How Buckminster Fuller's geodesic dome for New York might have looked. The segments for the 1.6km high, 3km wide dome were to be flown into place by helicopters.

own weather, which meant no long, snow-clogged winters, and rain only when it was needed.

These were big ideas from a man with an enormous intellectual capacity – he went on to become president of Mensa, the international organisation for people with high IQs. There was personal motivation, steeped in tragedy, too, behind Buckminster Fuller's drive to make the world a better place. At just 4 years of age, his daughter had lost her life to an illness caused by poor housing. That this could happen in one of the world's most advanced cities, in the modern day and age – 1929 – was incomprehensible to Fuller. So he set to work on a variety of inventions that pushed the frontiers of engineering. His plans for levitating cars and boats that were propelled by opening and closing cones above them were all part of the learning curve that led to the Manhattan dome project.

The timing for the development of geodesic domes was perfect. Although originally patented in the 1920s by German designer Walther Bauerfeld, Fuller believed he could improve upon them. They were ingenious, they were strong, but Fuller thought that with some changes they could do more for mankind than simply house exhibits, which was Bauerfeld's original purpose. Examining the mathematical and engineering concepts behind the domes, in 1948 he discovered the mathematical formula for the closest packing of spheres. The theory was this. Geodesic domes can cover extensive areas without internal supports, and accordingly can contain more space with less material than other structures. That means they are relatively cheap too. Moreover, the bigger they are, the stronger they become, so they can contain a lot of city. Finally, because they keep the elements out and the heat in, they can be made to adopt their own internal microclimate. Hot or cold, you can create the climate you want.

Casting himself as a 'comprehensive anticipatory design scientist' – comprehensive meaning he wanted to benefit the maximum number of people using the minimum amount of resources – domes could fulfil a useful environmental function; improving housing so that others would not have to suffer the death of a child. Fuller looked at inhospitable parts of the globe such as Africa and Antartica and could see a use for the domes. One idea was for a tetrahedral city that would float in Tokyo Bay and house a million people. But it was the Manhattan Dome that captured America's imagination.

But could it be done? An early classroom-based attempt to construct a test dome out of Venetian blinds failed when the dome fell in on itself almost as soon as it was completed, causing cynics to name it the 'flopahedron'. In response, Fuller argued that the collapse was intentional – he wanted to understand the critical point to which he could go whilst in a safe environment. Then came a success: the Ford Motor Company commissioned the first commercial geodesic dome from Fuller in 1953, spanning 93ft at a weight of just 8.5 tons. With a successful project under his belt, demand for geodesic domes suddenly rocketed.

Plans for Manhattan's dome progressed. True, at $200 million it would be costly, but the 4D concept of lifetime's use came into play: over time, living inside a dome that was constructed out of fewer resources than a traditional building would save cash. The climate could also be efficiently managed. Indeed in today's terminology, geodesic domes were green, not least because the dome would pay for itself within ten years in the cash saved from clearing away New York's snow alone. Eventually the site was located, spanning '50 blocks of the upper Manhattan skyscraper city' from the East River to the Hudson at 42nd Street, and north and south from 62nd Street to 22nd Street. Sixteen large Sikorsky helicopters would fly all the segments into position for the 1.6km high, 3km wide dome. It would be constructed within three months.

But construction never took place. The political will was lacking and so was broad residential and commercial support. It's easy to get excited by a domed city when you live miles away, but less so if someone wants to encase your home or business in one. For many people, the whole concept was too futuristic. And although he went on to be nominated for the 1969 Nobel Pace Prize, Fuller's inventions sometimes lacked credibility. His Dymaxion Vehicles, three-wheeled cars that could do a 180° turn in a single swift maneouvre, which was handy for getting in a parking space but little else, caused such traffic chaos that residents pleaded with him to keep off the roads at peak times. Odder still, Fuller claimed that dolphins evolved from humans, which wasn't entirely the prevailing orthodoxy, and that, as captain of 'Spaceship Earth' he would care for everyone on board, combining the role with that of something he called Guinea Pig B, just one of the nicknames he liked to be known by. His Synergetic Geometry system, based on 60° angles rather than the traditional 90°, failed to catch on too. But, most crucially, in addition to the $200 million bill and the resistance of Manhattan residents to seeing their homes encased, geodesic domes had flaws.

For a start, they weren't entirely waterproof, which while fine for encasing cities, didn't do much for the argument that they could control the climate. Next it proved difficult to subdivide the interior. Much of the available space was at a high level, which was impractical or expensive to use fully for other purposes. Even Fuller's own dome, one he had built as a home and where he lived for some time, leaked. And when builders set to work on repairs, they ended up burning down the whole building.

In the end, the project never progressed beyond Fuller's imagination and the drawing board. It did result in acres of positive press coverage and increased demand for domes around the world. The Pentagon invited Fuller to design protective housing for radar equipment while Soviet President Khrushchev ordered a dome for a Moscow fair and wanted Fuller to teach engineering to Soviet engineers. And although the Manhattan domed city didn't come to pass – or at least it hasn't yet – the prevalence of geodesic domes today is just one of Richard Buckminster Fuller's considerable scientific, environmental and social legacies.

EXPLODING TRAFFIC LIGHTS

Today, London pavements are clogged with tourists photographing Big Ben or waiting at the modern traffic lights to cross the road for a better look at the famous landmark. But most of those who linger momentarily at the corner of Bridge Street and Whitehall are oblivious to the tiny blue plaque on the wall above and behind them. It recalls the almost forgotten historic importance of the spot where they stand. For this is the sight of the world's very first set of traffic lights, erected way back in 1868 – a date when the Houses of Parliament across the street were only just receiving their finishing touches.

Traffic lights have, of course, become a norm of modern-day life and a crucial part of any town-planner's armoury. Love them or loathe them, they save lives and help us manage our traffic. Yet they could have been commonplace fifty years before they began to appear routinely at busy junctions. Sadly, a chance tragedy would kill off John Peake Knight's revolutionary idea before it was properly put to the test. Knight's gas-fired traffic light was to prove simply too ahead its time, falling foul of the technology then available to him and pressure from the Victorian health and safety lobby. Incredibly, the unlucky demise of his gadget meant that no one else would

POLICE NOTICE.

STREET CROSSING SIGNALS.
BRIDGE STREET, NEW PALACE YARD.

CAUTION.

The Semaphore Arms lowered, and by Night with a Green Light.

STOP.

The Semaphore Arms extended, and by Night with a Red Light.

By the Signal "CAUTION," all persons in charge of Vehicles and Horses are warned to pass over the Crossing with care, and due regard to the safety of Foot Passengers.

The Signal "STOP," will only be displayed when it is necessary that Vehicles and Horses shall be actually stopped on each side of the Crossing, to allow the passage of Persons on Foot; notice being thus given to all persons in charge of Vehicles and Horses to stop clear of the Crossing.

METROPOLITAN POLICE OFFICE,
December 10th, 1868.

RICHARD MAYNE,

Commissioner of Police of the Metropolis.

Above: Police poster from 1868 warning Londoners of the new gas-powered traffic lights which briefly appeared in the capital, before a tragedy put an end to the experiment.

Right: The blue plaque at the corner of Bridge Street and Whitehall.

try and introduce a traffic light to any street corner anywhere in the world until well into the twentieth century.

Even before the age of the petrol engine, however, some sort of traffic flow device was badly needed. In Victorian London the roads were clogged with horse-drawn carriages and arguably almost as dangerous as they are in today's age of the motor car. During 1866 a total of 102 people were killed on the capital's roads. In 2009 the figure was 184.

J.P. Knight, a Nottinghamshire railway engineer, believed he had the answer to the growing death toll on the roads. Born in 1828, he left school at just 12 and worked his way up to become traffic superintendent on the South Eastern Railway. His obituary later told of how he was particularly good at 'managing the crowds on Derby Day' and was instrumental in the adoption of new braking systems for trains, as well as emergency bell-pulls for lone female travellers in train carriages. Concerned too about the state of traffic management on the roads he wrote to the Home Secretary recommending one way streets. His letter read: 'Narrow streets, where two vehicles cannot well pass each other should be used for traffic in one direction only, so as to prevent vehicles meeting.'

Then, in December 1868, two years after his idea for traffic signals in central London went before a Parliamentary Select Committee, the authorities gave the go ahead. *The Express* reported at the time:

> The regulation of the street traffic of the metropolis, the difficulties of which have been so often commented upon, seems likely now to receive an important auxiliary. In the middle of the road between Bridge Street and Great George Street, Westminster, Messrs Saxby and Farmer, the well known railway signalling engineers, have erected a column 20 feet high, with a spacious gas lamp near the top, the design of which is the application of a semaphore signal to the public streets at points where foot passengers have hitherto depended for their protection on the arm and gesticulation of a policeman – often a very inadequate defence against accident.

As this indicated, the design of the Knight's traffic lights, which relied heavily on train signals, featured two semaphore style arms. When they were raised it meant traffic should stop. When lowered that they could proceed with caution. At night-time red and green gas lights would also be illuminated thanks to power from a gas pipe running up the middle of the post. The whole thing was to be operated by a policeman with a lever with the arms raised to stop traffic for thirty seconds in every five minutes. Wags joked that the Home Office was keen on the idea because it would protect the lives of MPs and officials going to and fro Whitehall.

On 10 December the first set of traffic lights were installed at the junction where traffic poured off Westminster Bridge and met the throng coming down Whitehall. Made from cast iron and painted green and gold, the typically grand Victorian devices were even topped with a pineapple finial. A poster, now in the Metropolitan Police archive, was put up to inform travellers about the new device. It read:

> By the signal 'caution', all persons in charge of vehicles and horses are warned to pass over the crossing with care and due regard to the safety of foot passengers. The signal 'stop' will only be displayed when it is necessary that vehicles and horses shall be actually stopped on each side of the crossing, to allow the passage of persons on foot; notice being thus given to all persons in charge of vehicles and horses to stop clear of the crossing.
>
> Proclamation of Richard Mayne, London Police Commissioner

In fact, a staggering 10,000 of these leaflets and posters were put up all over London.

The Express enthused: 'A more difficult crossing place could scarcely be mentioned and should the anticipations of the inventor be realised similar structures will, no doubt be speedily erected in many other parts of the metropolis'. Indeed, plans were soon afoot to install the signals on other busy roads in the capital, like Fleet Street. Then, without warning, on 2 January, the traffic light exploded. The policeman operating it was badly burned. Some reports suggest he later died. The injuries were certainly bad enough for the traffic light to be quickly taken down and the idea was completely dropped. The gas which fuelled the lights had proved too dangerous.

Ironically, Knight also worked on the first electrically lit train carriages. And only with the advent of electricity was a similar idea to his traffic lights eventually revived in 1912 by the aptly named Lester Wire, a police officer in America's Salt Lake City. It wasn't until 1926 that London was again to get its next set of traffic lights – put up at the junction of St James's Street and Piccadilly. Even this still had to be operated by a policeman before automated lights finally came in on 14 March 1932.

Knight himself died in 1886 following a stroke. Some 2,000 people turned out for his funeral procession and a wreath was sent by the Prince of Wales. He was just 58. Who knows? Had the genius of his invention been properly recognised, tested and perfected the premature deaths of many more might have been saved.

THE STEAM-POWERED PASSENGER CARRIAGE

For an enterprising soldier eager to advance, little beats inventing an improved way of killing people. So when Nicholas Cugnot designed a new rifle at the very time the French General of Artillery, Jean Baptiste Vaquette de Gribeauval, was working on his own system for standardising weapons, he endeared himself into the higher echelons of the military. Cugnot's rifle was soon part of the French army's arsenal and, impressed with the young engineer, Gribeauval found further work for him. It was to lead to the very first powered car, the very first car accident and, quite possibly, the very first arrest for dangerous driving. The steam-powered car was on the way, and record breaking though it was, it wouldn't stand the test of time.

Having already revolutionised French weaponry, in the 1760s General Gribeauval set to work improving the way munitions were moved. Horses, the traditional form of transport in the eighteenth century, had been hauling cannons and guns since the dawn of warfare. But, at a time when British industrialist James Watt was improving the steam engine, this was the new age of transport. With an engineer in his ranks who had already proved his talent for military design, the well-connected Gribeauval, through the Duke of Choiseul, minister of war, arranged funding from the defence depart-

Nicholas Cugnot's *fardier à vapeur* on display at the *Musée des Arts et Métiers* in Paris. His 'boiler on wheels' made its first journey in 1769.

ment and charged Cugnot with inventing a new machine using an external combustion engine – steam.

By 1769 Cugnot had patented the world's first 'motorised carriage', basically a boiler on three wheels that could reach speeds of 2.5 mph by converting the lateral energy of steam into rotational energy to power wheels. His breakthrough was placing the engine over the front wheel, and then using two pistons attached to each wheel to propel the vehicle. With a driver on board to steer the unstable, weighty vehicle in as much of a straight line as possible (changing direction was a major operation), as well as stoke up the boiler, the car was born. But there were problems. With the boiler at the front, it had a tendency to tip onto its nose unless counterweighted by a canon at the back – although as its mission was to tow 5 tons of armament this wasn't necessarily a bad thing. Of more concern, it could only produce enough steam to run for fifteen minutes at a time. And there wasn't a great deal of space to carry spare fuel or water. In short, this was a vehicle that most people could out-walk. But it was a start.

Cugnot tried again. The first ever mechanically propelled vehicle may have been cumbersome, more tractor than car, but it worked. What if he could improve speed and reach and make the cabin larger so that it could take passengers? Within a year, with his colleague, army mechanic Michel Brezin, and with the support once more of the duke and the general, Cugnot's next steam vehicle accommodated four people. As with the earlier version, two pistons connected by a rocking beam were synchronised, so when atmospheric pressure forced one down, the other went up, creating reciprocating motion through the axle to turn the wheels. But this time, the pistons could be moved without condensation from the high-pressure steam engine, thus increasing efficiency. And the vehicle was in two parts – at the front, a copper boiler with a furnace inside and two small chimneys above, and at the back the carriage on two wheels for the passengers and a seat at the front for the driver.

The duke and the general, together with other French military dignitaries watching the trial, were impressed. No longer did the driver have to stop every fifteen minutes to add water, then sit around waiting for it to boil. The vehicle, which could now travel for an hour and a quarter without a break, was still harder to operate than a horse-drawn carriage and, with the invention of brakes still years away, was all but impossible to stop. Nevertheless, approval was granted for further development.

Sadly, in his quest for greater efficiency, brakes weren't top of the list for Cugnot. One day in 1771, while trundling along at a speed of up to 3 mph in his mark two model *fardier* (French for dray, as the car was known), he became both victim, and perpetuator, of the first car accident when, unable to brake, he steered the steamer into a wall. Some reports suggest that this was just the first in a series of accidents. The vehicle also overturned while attempting to turn a corner in the centre of Paris. Tired of picking him out of hedge and ditch, the police eventually charged Cugnot with dangerous driving, although no court records have been found to prove this slur and the earliest recording of the accident is in Cugnot's 1804 obituary. Quite coincidentally, the end of Cugnot's motoring career came not long after this collision. His military and political masters fell out of favour and Cugnot was pensioned off.

Self-propelled vehicles, however, continued to steam ahead. Within a few years, French engineer Onésiphore Pecqueur invented the differential gear, which allowed power to be transferred more efficiently to car wheels. In America, steamboats began to stay afloat, a dramatic improvement on the very first one in 1763, when William Henry's aqua invention sank. And by the early 1820s, steam-powered stagecoaches taught Britons how small the country really was. In the fine spirit of British progress, they were quickly outlawed because of the noise and danger.

If Nicholas Cugnot was the first person to get passengers moving, Father Ferdinand Verbiest, a Flemish mathematician and Jesuit missionary who moved to China to spread the good word, actually designed the first self-propelled steam car – but was unable to progress further than building a scale

Louis Ross racing a Stanley Steamer car, Florida, 1903.

model. His place in history is therefore as the creator of the very first powered toy car – more than a hundred years before Cugnot invented it for real. Somewhere between 1671 and 1679, Verbiest used steam to power a small turbine to turn the wheels of a vehicle. This work of genius he gave to the Chinese emperor, whom he had much to thank for. Jesuit missionaries in China didn't always go down terribly well and, a few years earlier, after being found guilty of teaching a false religion, Verbiest had been sentenced to death by a thousand cuts, China's most tortuous method of execution. Fate intervened when an earthquake destroyed the execution site and, as the authorities took this to be an omen, Verbiest instead ended up working for the emperor on redesigning the Chinese calendar (he ordered them to take out a month) and designing toy cars. When the steam vehicle powered across the floor, turned around and shuddered back to its owner, the court was delighted. The omen had been a good one.

However, technology inevitably put paid to the external combustion engine in cars. Steam vehicles, being heavy, were, as Cugnot discovered, often slow. Constantly getting water to the boiler was trouble. Extra water, lots of it, has to be carried or a condenser fitted, thus adding more weight. The engines were also noisy. When, in 1889, Karl Benz entered the fray with an internal combustion engine powered by fossil fuel, for a while it looked like the battle for vehicular supremacy could go either way. But the greatest steam car of all, the Stanley Steamer, still had something to prove. Destroying the misconception that steam cars are always slow, in 1906 driver Fred Marriott, at the wheel of his Stanley Steamer, reached 127.659 mph (205.5 km/h); beating four petrol-powered cars to win the Dewer Trophy speed event and grab what is still the land speed record for a steam powered vehicle. It was the last major triumph for steam. Although petrol and steam cars were neck and neck in the race to become de facto standard, by the 1920s, the quieter, less troublesome internal combustion engine had won the battle and the steam ran out of the external combustion adventure.

FLYING CARS

Airborne automobiles were once the future; so thought motoring icon Henry Ford. His mass production methods, which led to the famous Model T, heralded the expansion of car ownership in America in the first decades of the twentieth century. So people paid attention when, in 1940, he said: 'Mark my word. A combination airplane and motor car is coming. You may smile. But it will come.' In 2010 the USA alone had 136 million car owners and 187,000 civil aircraft. But the idea of combining the two is still mere fantasy. There isn't a flying car in every other suburban driveway and the skies above the world's major cities are not filled with winged private vehicles. It hasn't stopped a string of idealistic enthusiasts working on machines that you could use both on the highway and in the air. None have yet cracked the mass market as Ford expected, but he was certainly right in one sense: flying cars are perfectly feasible.

Three years before Ford made his prediction, one Waldo Dean Waterman was already leading the way. His three-wheeled Arrowbile was built around a six-cylinder Studebaker car engine. It had no tail and the engine was placed at the rear of the fuselage along with the craft's propeller. To turn it from a car into a plane you essentially fixed the wings (with a span

Moulton Taylor's Aerocar, first developed in 1949, never took off commercially.

of 38ft) to the top when you wanted to fly. 'Add wings and it smoothly takes to the airways', ran the advertising copy. In August 1937 *Life Magazine* reported that the Arrowbile, later called the Aerobile, could fly at a speed of 120 mph, carry two passengers a distance of 350 miles and was yours for $3,000. On the roads it could do a sprightly 56 mph. It could certainly fly and the Studebaker company ordered five. When it came down to it though, Waterman's figures didn't add up. His car-cum-plane turned out to be more expensive to make than it was commercially viable to sell and he didn't have the invest-ment needed to see the plan through.

In 1946 the baton was taken up by Robert Edison Fulton Jr. who, during the war, had invented a simulator-style machine for training aerial gunners. Flying himself across the US in a light plane for his work, he was frustrated at being stranded at airfields miles from the local town, unable to find a taxi or lift. He wondered why he couldn't just take his plane downtown. Despite having little expertise on aircraft design itself, Fulton nevertheless managed to build what he called the Airphibian. It was a simple but effective design, featuring a set of fabric wings and a tail that could be attached to the body of the car. Once removed you could drive from the airport back to your home and park the car in your garage. The transformation from plane to car, or vice versa, could all be done in five min-utes flat. The Airphibian had a car's accelerator and brake pedal which doubled up as the plane's rudders in the sky. It could fly a distance of 400 miles, reaching speeds of 110 mph in the air and 55 mph on the ground. In 1950 the four-wheeled Fulton Airphibian became the first flying car to be certified for production by the US government's flying authority, the Civil Aeronautics Administration. The organisation itself even ordered ten of them at a cost of $7,500. Again money was the problem. The machine had been required to undergo rigorous tests which meant Fulton was fast running out of cash and needed to bring in an investor. Disagreements with his backer meant Fulton eventually sold his own investment, abandon-ing the Airphibian altogether. Instead he went on to work on a much more successful device called the Skyhook, an aerial rescue system for retrieving spies and soldiers from behind enemy lines, used by the US military for some thirty years.

Another flying car builder, also American, was a former missile tester called Moulton Taylor. He'd seen Fulton's car in 1946 and reckoned he could make a more practical version. His Aerocar, which did similar speeds to the Airphibian, had folding wings which were designed to be towed behind the car when it was on the road. After raising $50,000 to build and test a prototype in 1949, it took Taylor another seven years before he got his government certification. Finally he was ready to sell them – at $15,000 a time. In 1958 he said: 'My dream is to look up and see the sky black with Aerocars – and I'm sure that will happen someday.' A company called Ling-Temco-Vought seemed to be on board too, promising to go into full scale production if Taylor could rustle up 500 orders. When he only got half that number the company pulled the plug.

Taylor, who had given up a job in a military lab, was dismayed but not beaten. He went on refining his design throughout the 1960s, but without a factory in which to build his cars he struggled to produce cheap enough vehicles to meet the demand he felt existed. Then, in 1970, there was an ironic twist to the tale. Ford, the company whose founder had prophesied the rise of the flying car thirty years before, got in touch. The company's researchers reckoned they could sell 25,000 models of the Aerocar a year. For a moment the machine appeared to have a new lease of life. But further research showed that simply to comply with tougher road and air regulation would cost some $400 million. Ford baulked at the sum. In the end only six Aerocars were made.

Shortly afterwards, in the early 1970s, there was another bizarre attempt at a flying car which ended in spectacular failure. Engineer Henry Smolinski planned to go into production with his combined car and plane, the AVE Mizar, in 1974. On 11 September 1973 during a trial flight, the contraption came apart, killing Smolinski and the craft's pilot. Since then there have been many more flying cars. Many bold, clever and futuristic prototypes have emerged. But large-volume sales are still elusive.

One of the problems with the first flying cars was that they were still pretty expensive for the man or woman in the street. Critics pointed to the fact that they also looked pretty ugly for something that was still a luxury item. Lighter engines and new materials now appear to make flying cars a better financial and technological bet. But the main obstacle probably has never been the technology, so much as the general unease from big manufacturers and, above all, the authorities. The idea of managing millions of small aircraft whizzing about the already crowded skies above the world's major conurbations must seem daunting for any government. With that in mind, any company that looks to develop a flying car faces not only huge development costs but a bureaucratic health and safety minefield. Flying cars might take pressure off our congested roads, but crashes could be pretty drastic affairs. And would they really free things up? Given that drivers would probably be forced to fly in narrow flight corridors, terrible tailbacks in the air would surely become commonplace too.

THE ATOMIC AUTOMOBILE

Imagine being able to clock up 5,000 miles in a state of the art vehicle without ever having to refuel. That's the idyllic vision of carefree motoring that the world's first nuclear powered car, the Ford Nucleon, promised to deliver. But it won't be coming to a showroom near you any time soon. And that might be just as well. Just consider the insurance payments on a car that could, potentially, lay waste to the very streets you're driving through.

Back in 1958, however, the idea of an atomic car promised to open up a brave new world of road transport. Scientists at Ford envisaged a time when mini nuclear reactors would be commonplace in everyday machines. In its glossy brochure to launch the idea of the Nucleon – surely the twentieth century's boldest car design – the company promised a 'glimpse into an atomic-powered future.' Its designers explained that such a car would, in time, be possible, because 'the present bulkiness and weight of nuclear reactors and attendant shielding will some day be reduced'.

While a full size Nucleon was never built, the firm did go as far as constructing a 3/8 scale model, complete with classic 1950s tailfin styling, as well as producing detailed publicity materials outlining how it would work. Its designers intended

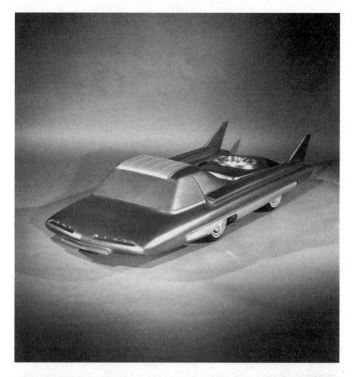

A scale model of the Ford Nucleon, a concept car from the 1950s, which was to be powered by a mini nuclear reactor.

to do away with the internal combustion engine. Instead the vehicle would be built with a power capsule containing a radioactive core element – sitting cosily in the car's boot. They explained that 'the drive train would be part of the power package and electronic torque converters might take the place of the drive train now used'.

Able to travel thousands of miles without refuelling, cars like the Nucleon would make fossil fuels obsolete. When the Nucleon did finally need re-fuelling, you'd simply take it to a special recharging station. Owners could even order different sized reactors depending on how many miles you were likely to travel. The brochure continued breezily: 'The passenger compartment of the Nucleon features a one-piece, pillar-less windshield and compound rear window and is topped by a cantilever roof.' There were other potential benefits, apart from not having to fill up. The car wouldn't need an ignition, it would probably be quieter to run than a traditional car engine and there would be no nasty emissions. Unless, of course, there was a crash! Not a word is included about the possible risks to the driver and passengers of a radioactive leak, or the potential havoc caused by a nuclear meltdown on the High Street after a prang.

The concept of the Nucleon was born at a time when the popular belief was that nuclear energy could be utilised to solve many of society's problems and power almost anything. The era was dubbed the 'atomic age' and while many were awed by the destructive power of nuclear weapons, they also saw this incredible energy source as heralding a new industrial revolution. In 1954 Lewis Strauss, then chairman of the United States Atomic Energy Commission, predicted that nuclear power offered the prospect of 'electrical energy too cheap to meter'. The first full-scale commercial nuclear power station opened at Calder Hall in Britain in 1956.

Within a few years, however, the public began to be more wary. Concern grew about the limitations and dangers of nuclear devices, with a number of accidents at nuclear power plants. In America the Three Mile Island accident in 1979 was also an important milestone. The growing sense of unease may have been one reason why Ford didn't progress with its research into nuclear-powered cars. It's easy to see why. Aside from the safety aspects, the technological challenge and cost of developing the car would have been huge. There also seems to be no thought given to what to do with all the nuclear waste these cars would have produced.

While Ford were toying with atomic-powered cars, experiments were being made to use nuclear power in other methods of transport. Both the US and Soviet governments experimented with nuclear-powered aircraft in the 1950s. Such planes would have been able to stay in the air for very long periods of time. Improved ballistic missiles eventually made the idea redundant; though even today, some scientists are debating the merits of nuclear-powered aircraft as one of the answers to the problem of global warming. Nuclear-powered submarines, meanwhile, have been around since the USS *Nautilus* put to sea in 1955.

While military applications of nuclear fuel are often considered safe enough to be used, it's not now likely that governments would allow thousands of nuclear-powered private machines to ply the highways and byways of the land. Though perhaps they would if vehicles all travelled on an automated road, controlled by a computer; another idea that has been knocking around since the 1950s. It's easy to laugh at the notion, but, with oil production soon set to peak, how long does the combustion engine really have left? While the motor industry is working on cars powered by alternative fuel sources from electricity to hydrogen, the resulting machines have yet to make a significant dent in the consumption of petrol and diesel.

In the future we may certainly need the sort of revolutionary idea embodied in the Nucleon, whose designers boldly stated in their 1958 brochure that they refused to 'admit that a thing cannot be done simply because it has not been done'.

21

THE X-RAY SHOE-FITTING MACHINE

When German physicist Wilhelm Röntgen's 1895 discovery of X-rays (then known as Röntgen rays) won the 1901 Nobel Physics Prize, Thomas Edison, the inventor of the concrete house and a man always eager to build on scientific advancements, paid careful attention. Here was a concept worth developing. Within six months, the Edison X-ray tube was ready for market, and fluoroscopy – the use of X-rays – began entertaining crowds as they gaped inside their own bodies and at the bones of hands, feet and head. Röntgen's X-rays and Edison's fluoroscope paved the way for an invention that was to change the way people were fitted for shoes, with lethal effects for shop assistants, and not much good for the very feet they were designed to help either.

At the start, and indeed for decades following, the public greeted X-rays with awe. To be able to see into the body fascinated even the most sceptical, as one man who complained that the fluorescent screen was nothing but a sheet of ground glass discovered to his embarrassment after being hauled onto the stage at its 1896 launch. Edison, deploying his customary showmanship, forced the man to hold up his wrist until he saw a hole through it, converting him instantly to the wonders of fluoroscopy. At the time, with no reason for concerns

An X-ray shoe-fitting machine from 1927, used in the Boyce & Lewis store in Washington D.C., is now on display at the National Museum of Health and Medicine in the city.

about the affect of the machines on health, everyone wanted a shot of their insides – one happy patient benefiting from a medical X-ray on the operating table on the very day of Edison's public exhibition. Soon the fluoroscope was enjoying commercial success in many forms; by the turn of the century being used to investigate complaints as complex as impotency and insanity.

The fluroscope had viewfinders for the shop assistant, parent and child.

Very quickly, however, fluoroscopy's darker side began to emerge. In 1900, Clarence Dally, Edison's colleague who had worked most closely on the fluoroscope, began to suffer skin lesions on the very area of his arm that he had habitually stuck into a machine during experiments. When skin grafts failed, first a hand was amputated and eventually both arms. Even this was insufficient to save him. Before long, cancer had taken him. A shocked Edison, who had almost lost his own eyesight, stopped working with fluoroscopes and began to warn of the dangers. 'Don't talk to me about X-rays,' he said. 'I am afraid of them … and I don't want to monkey with them.'

The concerns of an eminent, if eccentric, inventor, did little to prevent the fluoroscope's adaptation in ingenious ways though. Some products were undoubtedly useful: the work of Röntgen, Edison and later Marie and Pierre Curie created the basis of the science of radiotherapy that remains in use today. Other ideas were little more than gimmicks, trifling new products or marketing wheezes. In 1924, in Milwaukee, USA, resisting protests from the Radiological Society of North America ('it lowers the profession of radiology'), surgical supplies salesman Clarence Karrer marketed a shoe-fitting fluoroscope. But failing to progress a patent application, Boston doctor Jacob Lowe beat him to the legitimate claim of the invention. Lowe, who had been working on a device since returning soldiers from the Great War turned up at his surgery with foot injuries, developed a machine that could X-ray feet without the need for boots to be removed. He applied for a patent in 1919, emphasising the medical advantage:

> With this apparatus in his shop, a shoe merchant can positively assure his customers that they need never wear ill-fitting boots and shoes; that parents can visually assure themselves as to whether they are buying shoes for their boys and girls which will not injure and deform the sensitive bone joints.

Quite coincidentally, by the time Lowe's US patent was granted, in the UK, the Pedescope Company of St Albans had bagged their own patent. It had been 'in continuous daily use throughout the British empire for five years', they declared. So the claims to the first use of this ultimately dangerous invention are ambiguous.

By the end of the decade, the Pedescope Company and, in the US, the Adrian X-Ray Company of Milwaukee, were selling the devices widely. Almost every enterprising shoe shop had one: a large wooden box, nicely polished and presented, into which customers would place one foot at a time to see how snugly new footwear fitted. The length of the exposure varied depending on the size of the foot - the highest intensity for men, the middle one for women and the lowest for children. A timer set exposure time from five to forty-five seconds, the most common dose being twenty seconds. Shop assistant, child and parent would gape through one of three viewfinders at the bones of the feet and the outline of the shoe, and a quick wiggle of the toes would reveal growing-room. The only shield between feet and tube was a miniscule aluminium filter. Shoe shops with pedescopes had a marketing advantage.

Reluctant children now looked forward to trying on shoes, and it was the sign of a responsible parent. Remember, yelled a press advertisement for the Adrian X-Ray machine: 'They'll need their feet all through life.'

Buoyed by success and the sound of ringing tills, marketing intensified. The Adrian X-Ray Company implored stores to 'place the machine in the most desirable location, [facing] the ladies' and children's departments by virtue of the heavier sales'. Press advertisements assured parents that: 'shoes that fit well, last longer.' And a well-shod foot was better for health; safer: 'The Adrian special shoe fitting machine has been awarded the famous Parent's Magazine Seal of Commendation … a symbol of safety and quality to millions of parents all over America.'

Any concerns about dangers were slow to develop. But towards the end of the 1940s, with the machines now deployed for more than twenty years, reports of skin and bone marrow damage and growth problems were emerging. After two atomic bombs brought an end to the Second World War, the effects of high radiation dosage were becoming recognised. In 1950, researchers Leon Lewis and Paul E. Caplan asked whether the shoe-fitting fluoroscope was hazardous. 'Physicians and health physicists everywhere are beginning to be concerned about every potential source of radiation … quite irrespective of the potentialities of the atomic bomb.' But even at this time, they said: 'The shoe-fitting fluoroscope is not an instrument with obviously hazardous potentialities. It has long been used and no direct clinical evidence of harm has yet been established.' So the machines continued to sell. By the middle of the century, more than 10,000 were in use in the US and a slightly smaller number in the UK, Canada and Europe.

Medical evidence about the detrimental effects did emerge. More than one shop assistant, used to putting their hands under the X-ray to pinch around the shoe to confirm the fitting, suffered hand lesions or dermatitis. Injury to reproductive organs from prolonged exposure was also reported; one shoe model had to have a leg amputated. Children were at greatest risk. Throughout the 1950s and '60s, US regulators gradually

restricted the advisory amount of radiation to which customers, and children in particular, could be legally exposed. Only
twelve doses per child per year, several states recommended
– which still allowed a lot of shoe fitting. Some precautions, however, backfired. When the American Conference
of Governmental Industrial Hygienists (ACGIH) suggested
a uniform set of standards in 1950, manufacturers met the
regulations and then boasted about it in a way that suggested
that pedescopes were government approved. Nonetheless,
by the 1960s, thirty-three states had banned the device altogether, while in others only a licensed physician could operate
them. Most shoe shops, being reluctant to hire a doctor to fit
shoes (and face parents wondering why a medic was suddenly
required to do something that had previously undertaken by
a teenage boy earning pocket money), removed the machines.

The UK lagged behind America. While US states fell over
themselves to protect customers, British regulators resisted. As
late as 1960, the Pedescope Company of St Albans was defend-
ing their machine against an attack in *New Scientist* magazine:
'The only real way of getting a higher standard of shoe fitting
is to let both the customer and the sales person *see* inside the
shoe the damage that is being done by a badly fitting shoe,'
wrote its representative. 'The difficulty of shoe-fitting being
a "blind" operation whose effect whether good or bad only
becomes apparent later on is completely overcome by the use
of a pedescope.'

Although some machines were still in use in the 1970s –
and one was in service in West Virginia as late as 1981 – the
tide was turning. What fifty years earlier had appeared to be
exciting, modern and infallible was, by then, a potential source
of harm, even death. But throughout that period, even more
remarkable radioactive inventions were on the market, including nuclear foods and drinks. Their origins were in a form of
radioactivity called radium. And it is that to which we turn
next.

22

THE CURE THAT KILLED

In the last decade of the nineteenth century, new discoveries in physics came thick and fast: Heinrich Hertz's high frequency radio waves in 1886; Wilhelm Röntgen's shortwave equivalent, X-rays, in 1895; and Henri Becquerel's identification of spontaneous radiation shortly afterwards. Then, in 1898, Marie and Pierre Curie discovered a new substance that emitted radiation. Coining the term 'radioactivity' they named the element radium – and unwittingly started a craze for radioactive gadgets, beverages and quack cures.

Once discovered, the Curies quickly set to work to truly understand how radium was constituted. From a shed at the school at which Pierre had a workshop, they examined the element intensively; the luminous substance emitting an intriguing blue glow that warmed Marie's heart in more ways than she realised at the time. It was joyous, she said, to go down to their shed at night and see 'from all sides the feebly luminous silhouettes' emanating within. The light – and the radiation – that radium emitted could be seen through the thickest substances. On the evening they jointly won the 1903 Nobel Physics Prize with Henri Becquerel, Pierre entertained their friends by producing a little tube from his waistcoat which, throbbing with a blue light, illuminated the small party of beaming faces. He wasn't aware of the implications – and, being run over by a horse-drawn carriage three years later, never lived to see it kill Marie – but the Curies had invented glow-in-the-dark products and they were to set the world alight.

The perennial puzzle of how to tell the time in the middle of the night was among those solved by the Luminous Materials Corporation, which from 1915 to 1926 produced radium from a mineral called carnotite. Young girls in their first jobs out of school mixed radium powder with glue and water and, with fine strokes from a camelhair brush, painted luminous hands on watches and clocks. When the brushes lost their fineness, which they did very quickly, the girls pointed them back into shape with their lips. The taste of radioactive paint aside, the work, while repetitive, could be fun. When one of the teenagers discovered that her handkerchief glowed in the dark after she'd blown her nose, it opened all kinds of possibilities for mischievous entertainment. By slapping the mixture on their teeth and faces, boyfriends could be frightened witless when the lights went out.

At least, to begin with, the girls had teeth that could be painted. This happy orthodontic pleasure wasn't to last long, as factory worker Grace Fryer was to discover. After spending three years painting luminous dials at the New Jersey factory, Grace's teeth were in trouble. Within two years of leaving, her health was in serious decline and, although she didn't yet know it, she was dying. Loss of her teeth was just the start of her problems. Soon bone decay in her mouth and back was diagnosed. Grace wasn't alone in such misfortune. Most of her seventy colleagues were by now also suffering grotesque illnesses. One woman's palate 'had eroded so that it opened into her nasal passages', said a report. 'Tests showed her to be radioactive.'

Five women launched legal action against the company, now known as the US Radium Corporation. In the face of a ruthless campaign of misinformation by a firm determined to resist each $250,000 claim, the judge found in favour of the workers, but awarded just $10,000 in immediate cash and a smaller annual amount for life. Those lives turned out to be woefully short. By 1928, thirteen women were dead, needlessly young, and scores were to follow. One body, that of Amelia Maggia, suffered the indignity of being exhumed for further medical evidence during the trial; her bones found to be 'still luminous with the radium she swallowed'. Investigations many years after the event showed that even the dust on the factory floor was radioactive and an experiment with cats revealed how their bones decayed from inhaling it. Ninety years after the radium girls first painted watch dials, radiation levels were still being assessed at the site.

In many respects, the women were fortunate to win their case against a company manufacturing a product that, contrary from being a health risk, was seen as positively health enhancing. Well before the factory girls' glowing teeth were alarming amorous boys, and even throughout the trial and well beyond, companies were adding zing to their products by including the wondrous ingredient. Hair tonic, sweets, toothpaste, toys, condoms, bread: all now with added radium. But no sector embraced radium more than medicine. Almost as soon as the Curies announced their discovery, pharmaceutical companies began to market radium ruthlessly. As an antiseptic cream, a malaria cure, the very thing you needed if you suffered from kidney or liver disease, indigestion, or even insanity; radium would help. All symptoms could be relieved, of course, by taking a radium bath. Arthritis or rheumatism sufferers were advised to hold their faces over a bowl of boiling water, add a dash of radium and inhale. Old, ugly or vain people had radium to thank for its skin-rejuvenating properties, which wiped years off them. And if all this sounded like quackery, any concerns, and there were few, were assuaged by the science: the correlation between radium inhalation and improvement in arthritis and rheumatism sufferers, for example, had been proven, erroneously as it turns out, as early as 1906.

The most popular product of all was 'liquid sunshine' (radioactive glowing water), launched in 1904 by Dr George F. Kunz and Dr William J. Morton, professors of electro-therapeutics at the New York postgraduate medical school. To illustrate a talk on how to make luminous drinks out of radium, the doctors had managed to acquire for the evening the world's biggest diamond, the Tiffany diamond. Placing a piece of radium behind the jewel, the room lit up. 'The light seen was like that of phosphorous,' said the New York Times. 'Afterwards the diamond was placed in a glass of water where it shone beautifully.' With 'liquid sunshine', the name the doctors admitted was chosen for its sales appeal, 'the whole interior of a patient could be lighted up'. Furthermore, the audience was told, radium may be the substance that gives the world's spring waters their curative powers. Even greater evidence was to follow. British scientist J.J. Thompson, who won the Nobel Physics Prize in 1906 for his discovery of the electron, had also found radioactivity in well water; the result of the presence of radium. Where radium (what today is known as radon) is present in the rocks over which water flows, the water was thought to be blessed with curative properties. A rush to the spas followed. Across America, health resorts boasted of the radium they possessed naturally in their springs. In Europe, Marie and Pierre Curie encouraged people to head to *ematorias* or *inhalatorias*, as spas were known, to soak up refreshing radium and radioactivity.

But not everyone could get to, or afford, spa treatment. As radioactive water has to be drunk quickly if the active ingredient, radium, is not to lose its powers, bottling water raised logistical issues. Instead, enterprising companies designed products that enabled customers to benefit from radium water at home. Amongst the first and most successful, in 1912 San Francisco company Revigator patented a 'radioactive water crock' and, although its $29.95 price was out of reach for many people, sold hundreds of thousands. The 'crock', a ceramic jug, contained radium which, when water was added and left overnight, would deliver 'the lost element of original freshness – radioactivity' to the water by breakfast. 'Fill jar every night. Drink freely. Average six or more glasses daily,' the packaging advised. It was 'a perpetual health spring in the home'.

The effects of the refreshing radioactive beverage could be long lasting. In 2010, brave chemists at St Mary's University, Maryland, USA, bought Revigator jars on eBay and found them emitting as much radiation as they did almost a hundred years ago.

The Revigator faced stiff competition from smaller, transportable, and, to the benefit of the radium-poor, much cheaper radium water products. Instead of water being placed in a jar, with the Thomas Cone, the Zimmer Emanator and the Radium Emanator, radium salts were dissolved in them. Americans especially were fond of such radioactive tipples, until stories began to emerge with alarming similarities to those of the luminous watch girls. Businessman Eben Byers, a millionaire steel magnate, knocked back 1,400 bottles of Bailey Radium Labs' Radithor in two years; his story culminating in a *Wall Street Journal* article headlined: 'The Radium Water Worked Fine Until His Jaw Came Off'. Only then did the public begin to take notice.

Yet the notion of radium as a medicine was deeply ingrained. Throughout the 1930s, sales of radium water slowed as scare stories, and evidence, mounted. Other new radium products, however, took their place: $150 would buy a 14-carat gold Radiendocrinator male pouch, or condom, in its own velvet-lined leather case. A nose cone respirator was cheaper, but less enjoyable. For soldiers ready for battle in the Second World War, the glow-in-the-dark radium products never really went away: radium tacks placed in barbed wire helped their comrades locate passage through, guns came with radium-lit sights for night-time aim, and in the trenches troops wore luminous wristwatches so they could see how much time they had to kill. Only by the middle of the twentieth century did the commercial market for radium fall away, the evidence by now too compelling to ignore; the deadly after-effects of radioactive poisoning too serious to risk. As a medicine, a toy, a gadget and a drink, radium had had its day.

23

THE 'CLOUDBUSTER'

Austrian psychoanalyst Wilhem Reich's plans to form clouds using a type of radiation he had identified, creating rain along the way, led to at least one bunch of happy farmers being relieved from drought. But 'cloudbusting' was just one of his ideas that led to his life's work – at one stage widely admired – eventually becoming much ridiculed.

Combining the looks of a space-age weapon with a rather cumbersome clothesline, Reich's 1940s' cloudbusters fired streams of potent 'orgone' energy that he claimed surrounded the earth. Any resulting rain was but a by-product of his main intention – which was to harness the sexual orgone energy and use it for the good of mankind. These were worthy aspirations from an intelligent scientist who ended his days certified and imprisoned, but whose 'surrealistic creations' (in the words of *Psychosomatic Medicine*) never managed to produce the results he expected. In keeping with psychoanalytic theory – an emerging discipline in which Reich was intimately involved – the root of the cloudbusters' failure lay in its inventor's childhood.

First Reich's mother committed suicide. Then his father, bereft at his wife's death, killed himself by standing for hours in cold water and contracting pneumonia. Reich's own attitude to the sex and the body was somewhat unusual. His mother had enjoyed a passionate affair with his tutor, which he'd worried about being forced to participate in. He'd seen the family maid have sex with her boyfriend and, in this case, volunteered to take part. In later life he would never go naked, even in his own company, and would generally wear his underpants in the shower to save any embarrassment.

With such intense interest in sex from an early age, Reich set out to dedicate his life to the study of the energy of orgasms. As one of a coterie of scientists working alongside Sigmund Freud, the psychoanalyst who postulated the concept of the 'id', 'ego', and 'superego' at the beginning of the twentieth century, his early career included extensive work linking the body and the mind in a way that is largely accepted today. Had he stopped there, he may have been remembered for his contribution to science. As it is, with the passing of the years, Reich became progressively more eccentric and his opinions increasingly extreme. By 1934, expelled from the International Psychoanalytic Association after arguing the link between good orgasms and good societies, he managed to upset all shades of European regime. Nazis, socialists, communists, all active in the interwar years, were all sexually repressed; the Great War was caused by the Kaiser's inadequate sex life. If only people would embrace sexual freedom, the world would be a better place. 'The masses of the people are endemically neurotic and sexually sick,' he told *Time* magazine later. His sex-political units, established in Austria to psychoanalyse the fervent political and economic environment (what he also called sex-economics), were understandably popular.

By now ridiculed in central Europe, Reich left for Norway where he 'discovered' 'bions' – blue organisms that destroyed bacteria and existed in a state somewhere between living and dead matter. This zombie life force caused skin to tan and emitted orgone radiation: 'the pure energy of life, the raw power of orgasm'. The press loved the story, but the scientific community scoffed. Reich migrated once more, this time heading for America where he began to build 'orgone accu-mulators', boxes the size of telephone booths in which patients would sit and feel the force of orgone. The aim was medical. Accumulators directed orgone energy – '*the* cosmic energy' as Reich was now defining it – to control disease.

With a fresh start in a new country, press opinion turned back in Reich's favour – for a while. Coming soon after his sensible studies of the link between body and mind, and with an American press that hadn't been subjected to his theories on sex-economics, orgone was initially received with appropriately respectful interest. *Science and Society* recommended his theories as something to watch. As work intensified in 1942, Reich moved his family to a large retreat in Maine that he renamed Orgonon in honour of his discovery. From now on, his inventions focused almost exclusively on orgone, but tests didn't always go to plan. 'Daddy put a radium needle in the big accumulator in the lab and everyone got sick,' said his son Peter. 'The lab closed, the mice died.'

But life was fun for a boy with an eccentric inventor for a dad, particularly when the cloudbuster rolled out of the lab. With its long, parallel tubes set in a frame and pointing at the sky, this was no toy. Anti-orgone, or Deadly Orgone Radiation (produced by atomic testing and UFOs), was changing the climate and the cloudbuster was the antidote. By absorbing orgone through hoses from the machine's pipe that dipped into water, precipitation could be produced. When fired correctly, streams of orgone would form clouds by creating a stronger orgone energy field than that in the atmosphere, from which the orgone would then be sucked down to earth and the resulting energy put to good use.

With the machine operational, Reich looked around the USA for a place to form clouds. Heading for arid Arizona, he packed a cloudbuster and son Peter into a truck and wound his way west. Aiming the apparatus at a clear blue sky, Peter fired. Not long after, it rained – at least according to Peter, who also used the cloudbuster to chase flying saucers of green and red disks: 'a cosmic adventure'.

For a man who was afraid of thunder and lightning, manipulating clouds was bravery indeed. And, while the cloudbuster was specifically intended to form clouds to bring orgone down to earth, a group of worried farmers from Maine hoped it could be the solution to a lengthy drought. On 6 July 1953, Reich was called in to save the state's blueberry crop. Local newspaper the *Bangor Daily News* reported that just hours after the cloudbuster went into action, the wind direction changed and rain fell. With blueberries saved, happy farmers paid Reich.

But it was the 'orgone energy accumulator' that was to do it for Reich. Believing that orgone-accumulating boxes could do much to reduce pain and disease, he began commercial production. About 250 accumulators went onto the market. Customers liked the idea of being cured of colds or impotency simply by sitting in a box and letting orgone accumulate, but the US Food and Drug Administration were far from impressed with its medical claims. A ban on distributing the product was issued, infuriating Reich. He was not to be stopped and sales continued. Even Albert Einstein is said to have had a go in an accumulator in 1941 when he paid Reich a visit. The Food and Drug administrators, though, remained unmoved. Reich's claims, they said, were fraudulent, and the case came to court. During his trial for contempt, Reich, who conducted his own defence, sent the judge all of his books, but the verdict, when it came, went against him. Reich, jailed for two years, his books destroyed and his reputation ruined, was finished. Showing no willingness to accept the authority of the court, he said he would carry on selling orgone accumulators regardless: thus dashing any hopes he may have harboured for a suspended sentence. Paranoia and delusions of grandeur were diagnosed as he went to prison. And it didn't get any better from there.

Aged 60 and just two months from release, a heart attack killed Reich, and with him died any hopes of establishing orgone as worthy of serious scientific study. He passed away believing that he was the victim of both a communist conspiracy and a cosmic war. Buried at Organon in a coffin he'd bought a year before his death, a replica of a cloudbuster stands by his graveside. His work is immortalised too in singer Kate Bush's album *Cloudbuster*, which also carries an illustration of the device.

24

THE BRAND NEW CONTINENT
OF ATLANTROPA

The idea that one could create an entirely new continent surely verges on the megalomaniacal, especially being conceived, as it was, by a German in an era which saw the rise of fascism. But the architect Herman Sorgel concluded that his colossal geopolitical scheme could solve the world's problems – or at least Europe's. He also believed that the technology existed to make his vision happen, even if it would take more than a century to complete.

Sorgel wasn't the only one. The German architect received many plaudits from other architects for his plan which involved building a massive dam across the Strait of Gibraltar, causing parts of the Mediterranean to dry up, creating a wholly new coastline and opening up vast new areas to inhabitation, cultivation and industry. Sorgel's ambition was to fuse Europe and Africa into a brave new land, which he named Atlantropa. Even Gene Rodenberry, the creator of the TV series *Star Trek*, was seemingly inspired by a similar notion. In his 1979 book he has Captain Kirk standing on a huge structure damming the Mediterranean to produce hydroelectric power, just like Sorgel.

Thankfully, the Nazis, who rose to power in the 1930s as Sorgel was busy sketching out his master plan, didn't seize on the idea. They were too busy eyeing up the territory that

Big Dam to Water Sahara

TURNING the Sahara Desert into blossoming farm land, with water drained from the Mediterranean Sea, is the ambitious project for which, Hermann Sorgel, German engineer, seeks international support. He proposes to dam the Strait of Gibraltar, and then cut a canal to flood portions of the Sahara below sea level. Evaporation from the inland lake thus formed would produce rain clouds and water a vast area, he maintains. By-products of the scheme would be hydroelectric power and new land reclaimed from the Mediterranean.

Above, artist's conception of the dam proposed for the Strait of Gibraltar and intended to help flood low parts of the Sahara Desert. At left, Herman Sorgel, originator of scheme, with model of Mediterranean

German architect Herman Sorgel working on plans for a dam spanning the Strait of Gibraltar; part of his dream to create the new continent of Atlantropa.

already existed to take Herman, a pacifist, too seriously. His works were banned by Hitler's government in 1942. Sorgel, who produced a staggering 1,000 publications on Atlantropa during his lifetime, thought his concept would help bring peace to Europe, which had already been ravaged by the First World War, as well as provide employment and ease the pressure from its expanding population. But many have branded Sorgel's Eurocentric views plain racist – especially his plans for Africa. In his world view, the supremacy of Europe was what mattered. He feared a future where the globe would be divided up into three huge competing blocks: America, Europe and Asia, with Europe potentially weaker than the other two.

So just what did Sorgel propose? Working as a sometime architect, writer and teacher he first started work on the Atlantropa idea in 1928 and refined his ideas over the years. The fulcrum of his plan, it emerged, was a massive dam stretching from Gibraltar across to Morocco, closing the narrow gap which separates Africa and Europe and links the Atlantic Ocean to the Mediterranean. The structure, which would be 18 miles long, and would take a million workers ten years to build, would be crowned with a 1,300ft-high decorative tower designed by fellow German architect Peter

Behrens. Another dam would be placed at the other end of the Mediterranean, where it is connected to the Black Sea at the narrow Dardanelles. Canals would be built beside the dams so that shipping could still pass through.

Sorgel said that these dams would reduce a natural inflow of water into the Mediterranean from the Atlantic and Black Sea. Not only would the dams produce huge amounts of hydroelectric power once this supply of water was cut off, the continuing evaporation in the Mediterranean would have the effect of gradually lowering the sea level. Within 100 years, said Sorgel, it would be 330ft lower. Eventually this would leave 220,000 square miles of virgin land ripe for exploitation and development. Sicily, for example would now be joined by a land bridge to mainland Italy, Corsica would be joined to Sardinia, while the Adriatic Sea would end up almost completely drained. Interestingly, the city of Venice would get its own special dam, effectively turning its famous lagoon into a lake.

Large parts of what Sorgel saw as an unproductive Africa would be flooded with another dam across the Congo River. It would involve moving two million inhabitants but Sorgel thought it was worth it. Meanwhile, the dunes of the Sahara could be reclaimed for farming. In July 1933 the American *Popular Science* magazine reported:

> Turning the Sahara Desert into blossoming farm land, with water drained from the Mediterranean sea, is the ambitious project for which Herman Sorgel, a German engineer, seeks international support. He proposes to dam the Strait of Gibraltar and then cut a canal to flood portions of the Sahara below sea level. Evaporation from the inland lake thus formed would produce rain clouds and water a vast area, he maintains.

Over the years Sorgel elaborated further. A bridge would be built between Tunisia and Sicily, easing travel between the two continents. The new dams, together with new power plants dotted around the Mediterranean, would produce a useful continent-wide power grid too. Sorgel even sought to collaborate with the German-Jewish architect Erich Mendelsohn over a new coastline for a potential future Jewish state in Palestine, years before Israel finally got statehood in 1948.

There were some big, practical problems with his ideas, not least that some of the Mediterranean's biggest ports would now be stranded inland. But in the 1930s many of his peers thought Sorgel was on to something. After all, that mammoth undertaking in America, the Hoover Dam, was completed in 1936. In addition, thousands of square miles of fertile land had already been reclaimed in the Netherlands when the Zuidersee, a shallow bay in the centre of the country, was dammed. Neither did many, back then, easily baulk at the idea of displacing millions of people, in what they saw as the interests of social and technological progress.

His plan does seem hopelessly grand. But Sorgel's fear that Europe would decline while America and Asia became more powerful has, to some extent, been borne out in the decades which have followed. Europe has certainly become more interdependent as he knew it would have to be. His fear that fossil fuels would begin to run out and that humankind would need renewable energy sources also seems forward thinking. As we know, Sorgel's ideas came to nothing, though interestingly the idea of building a bridge or tunnel between Spain and Morocco has recently been discussed between the two countries.

Sorgel, of course, was a fantasist. And despite spending hours coming up with great architectural schemes the only structures he ever built, according to reports, were a few houses. Even after the Second World War, his idea received plenty of attention and there was an Atlantropa Institute in Germany visited by foreign dignitaries keen to know more. For a man with such an exalted scheme, poor old Herman had a humble end. While cycling to give a lecture on the project he was run over by a car, and died on Christmas Day 1952. The idea of Atlantropa seems to have died with him.

25

A NATION BUILT ON SAND

In the 1949 Ealing comedy film *Passport to Pimlico*, a band of Londoners revel in the discovery that their part of the city isn't British at all but legally part of France. The movie's story taps into a deep desire among many to be free of the state they find themselves in. History has, of course, seen the rise of many virgin nations from the desire of peoples to be independent. Since 1990 the number of new states has grown by at least thirty. But, in a crowded world, it's been getting pretty difficult to start your own sovereign state from scratch. That hasn't stopped some people trying. A name for the more eccentric bids to state-building has even been coined: micronations.

While onlookers might find many of these attempts ridiculous, their proponents, often disgruntled groups or individuals, are often deadly serious. None more so than those who led the bid to found the Republic of Minerva; a nation constructed where there had once only been water. The main motivator behind it was Lithuanian-born Michael Oliver, a Jewish property mogul from Nevada. But this was not merely some late-night, bar-room fantasy. Oliver's dream of a libertarian state, with no taxes, did for a short time become a reality.

The nation's nickname, 'Land of the rising atoll,' gave a clue as to just how this country was to come about. It would be made from sand and gravel and be built on existing, unclaimed reefs that lay just a metre below the surface of the ocean. Located some 270 miles south west of Tonga, the Minerva reefs had been named after a whaling ship of the same name that had been wrecked there in 1829. The Republic of Minerva was also destined to flounder but not before a band of desperados had made serious practical attempts to set up an island nation that was ultimately designed to home 30,000 people, boast its own currency, constitution and capital.

In 1971, Oliver, backed by 2,000 Americans hoping to establish a rule-free capitalist utopia, arranged for barges loaded with sand to arrive from Australia. It was duly dumped and gradually the level of the island was built up, enough for a stone platform to be constructed. On 19 January 1972, a flag bearing a golden torch on a bright blue background was raised on this lonely spot in the middle of the Pacific Ocean. Coins were produced bearing the bust of the Roman goddess Minerva, along with the latitude and longitude of the proposed nation. Bizarrely the 10,500 Minerva dollar coins that were minted came with a value of 35 Minervan dollars. This did, at least, make them something of a valuable curiosity for collectors.

In February 1972, Morris C. Davis was elected as provisional president of the Republic of Minerva. Declaring the aims of the new country he said: 'People will be free to do as they damn well please. Nothing will be illegal so long it does not infringe on the rights of others.'

This certainly wasn't the sort of light-hearted episode that featured in *Passport to Pimlico*. It had got the neighbouring Tongans feeling uncomfortable and, in June 1972, the Tongan king decided to act. On 15 June, the *Tongan Government Gazette* reported that His Majesty King Tāufa'āhau Tupou IV had issued a proclamation that 'the islands, rocks, reefs, foreshores and waters lying within a radius of twelve miles' of the reefs were part of the Kingdom of Tonga. On 21 June, the king set out on board the royal yacht *Olovaba* for the reefs. With him were members of the Tonga Defence Force, a convict work detail and a four-piece brass band in order to enforce the claim. The republic's flag was lowered by the king who clambered onto the island himself. Quite an achievement given the fact that his highness was also his greatness; he weighed in at 350lbs.

But what about the other neighbouring nations? After all, the territory was actually a long way from Tonga. If there was land here for the taking wouldn't other countries fancy a piece of it too? As it turned out, the Tongans needn't have feared that the incident was about to spark a full-scale pan-Pacific conflict. A few months later The South Pacific Forum, made up of heads of government from the Pacific island states, recognised the Minerva Reefs as Tongan territory. For the Republic of Minerva, it was pretty much the end of the line as the nation slowly sank beneath the waves once more.

Oliver's attempts to start another nation certainly didn't end there, however. His Phoenix Foundation backed attempts in 1973 to declare a separate country in the island of Abaco, part of the Bahamas, and even trained a militia to take it by force. In 1980 the same organisation backed The New Hebrides Autonomy Movement, aiming to break free from the new nation of Vanuatu in the Pacific on the island of Espiritu Santo. It was quashed when the Vanuatu government organised for a battalion to occupy it.

Oliver's dream of a tax loathers' paradise was just one of scores of attempts to create micronations. One of the original examples was The Kingdom of Redonda. In 1865 a trader called Matthew Dowdy Shiell landed on the uninhabited Caribbean island and declared it his own kingdom. One hundred and fifty years later the island's ownership is still contested. There have been attempts closer to Britain, too. Sealand, founded by former British army major Paddy Roy Bates on 2 September 1967, was based on a derelict anti-aircraft tower 6 miles off the coast of Suffolk in the North Sea. In the past, British courts have ruled that they didn't have jurisdiction over the platform. His principality went as far as issuing passports, hurriedly revoked when crooks forged them – one emerged during the investigation of fashion designer Gianni Versace's death in 1997.

There aren't many free locations for anyone wanting to follow in the footsteps of these wannabe nation builders. Most of the world's landmasses are very much spoken for. But there is a part of the world that remains technically unclaimed – Marie Byrd Land in Antarctica, covering a massive 1,610,000 square kilometres. Until now the extreme climatic conditions mean no country has attempted to lay claim to this area as they have to other parts of the icy continent. But with global warming in full swing, now may be just the time to plant your flag.

26

THE DARIEN DEBACLE

In the late seventeenth century the Scots attempted to found a trading colony in Central America in an audacious bid to become a world power. Little did they know, this bold scheme would end in grizzly death and financial disaster. It was one of the factors which ultimately led to the Act of Union between Scotland and England in 1707, effectively bringing an end to hopes of Scottish independence for the next 300 years.

The Darien scheme was forged in the late 1690s by the Scotsman William Paterson, who had, ironically, been one of the founders of the Bank of England. Working in London around this time he'd met the explorer, buccaneer and surgeon Lionel Wafer who had just returned home after many years at sea. Wafer wrote a book called *A New Voyage and Description of the Isthmus of America*, describing his adventures in the region, and he inspired wealthy merchant Paterson with tales of an unconquered paradise ripe for the picking in the new world. According to Wafer the isthmus of Darien, in modern day Panama, was fertile, harboured friendly natives and was potentially the key to new trading routes to the spice-rich Far East. Here only some 50 miles separated the Atlantic and Pacific coasts. Establishing a route across this narrow strip of land would make long trips round the stormy tip of South America needless. Those that controlled Darien might dominate trade across the two oceans.

At this time Scotland and England shared a monarch, but Scotland had a separate parliament and could, technically, go its own way. But with famine and years of costly civil war, the country was suffering economically. Meanwhile the English, Portuguese and Spanish had been busy snapping up lucrative new territories around the globe. Already a successful merchant, Paterson came up with a plan. He would raise money for a colony called New Caledonia, a venture to make the Scots proud and, most importantly, rich again. In June 1695 it was sanctioned by the Scottish parliament.

Paterson was a good publicist and his enthusiasm for the loftily titled 'Company of Scotland Trading to Africa and The Indies' was infectious. While opposition from the rival East India Company saw off English and foreign investors, subscriptions from patriotic Scots worth £400,000 poured in to fund an expedition to Darien. People from even the lowliest levels of society signed up and many invested their life-savings. Estimates differ, but at least one fifth of Scotland's entire wealth was probably ploughed into the project.

In July 1698 Paterson was ready. Most of the 1,200 people who left with him didn't have any idea exactly where they were going until well into the journey. But, lured by the promise of 50 acres of land each, they embarked from Edinburgh in a fleet of five ships – the *St Andrew*, *Unicorn*, *Caledonia*, *Endeavour* and *Dolphin* – destined for a brave new world. The first aim was to start a settlement and get the colony on its feet with a view to establishing a new free trade port in Darien. In time, with trade busily passing through, the Scottish backers could then clean up with tidy commissions.

In November Paterson's group, minus seventy passengers who had died on the voyage, set foot on the shores of Darien. They lost no time building a fort to protect themselves, armed with fifty cannons. And a settlement, New Edinburgh, was founded. When news of their arrival reached Edinburgh on 25 March 1699, it was met with much celebration. The colonists

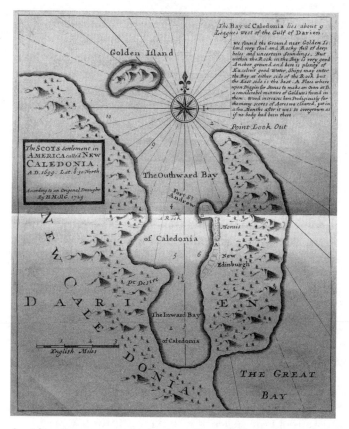

An eighteenth-century map showing the territory of Darien in Central America, the site of Scotland's failed colony.

began negotiating with the locals, and Paterson was among those who wrote back in positive terms about the state of the expedition that spring, perhaps not wanting to alarm the investors.

The optimism was not to last, however. The wet season soon arrived and tropical diseases like malaria wreaked havoc among the settlers; ten people were dying a day. Meanwhile, attempts to grow crops had proved tricky in a landscape thick with mangrove swamps and jungle. Settlers later recalled supplies riddled with maggots and one man told how his shoulders were covered in boils due to overwork. Attempts to trade for supplies with the local Kuna people also failed. Given the harsh climate it was not surprising that the natives were not impressed by the goods on offer: combs, mirrors, wigs and woollen clothes. The Scots soon resorted to hunting giant turtles to survive.

There was another problem. A mountain range stood in the way of any trading route that was to be pushed through to the Pacific, even if there had been enough settlers strong enough to build one. After eight months the decision was taken to abandon Darien. Of the 1,200 who had set out, only 300 of the original émigrés would survive to set foot in Scotland once again. Not knowing the fate of the first colony, a second fleet had left Scotland in August 1699 with another 1,302 settlers. It arrived in November to find Darien almost entirely deserted, bar a handful of original settlers who had decided to stay on. The new contingent began trying to rebuild things, without much luck. One of the eventual survivors, a Reverend Francis Borland, described conditions as 'pernicious, unwholesome and contagious'.

By now the Spanish, whose own new territories were dotted around Darien, were posing another threat. The English would be no help in this regard. As many embittered Scots would later relate, King William III (King William II of Scotland) wanted to drive the Spanish into the arms of the rival French and would not allow Darien any armed help from nearby Jamaica. English ports and ships were also forbidden to trade with the new Scottish colony. In 1700 some of the settlers at Darien made a pre-emptive strike against local Spanish forces, with some success, but soon the Spanish were back, besieging Fort St Andrew. Wracked by disease, the inhabitants were by now dying at the rate of sixteen a day. They held out for a month before surrendering and, in March 1700, those still alive set sail for home. Tragically, on the way back to Scotland, many more died from illness and shipwrecks, with just a handful eventually returning.

Some have suggested that if help had been forthcoming from the English, the Darien colony could have survived and even gone on to be successful. Whatever the reasons for failure, the company had lost a ruinous £232,884 and some 2,000 people had died. Politically, the calamity had left Scotland weaker rather than stronger. Paterson wrote a tell-all account of the Darien venture in a bid to justify the affair, but he himself became an exponent of the Act of Union; uniting Scotland and England's parliaments. As part of the agreement, the ailing Scottish economy would be bailed out. Paterson was eventually granted a pension of £18,000 but the whole debacle had come at a huge personal price. His own wife and son had perished in Darien.

Today the site of the Darien settlement near the Colombian border is still, by all accounts, a pretty wild place. Its Spanish name, Punta Escoces, or Scottish Point, recalls its ill-fated past. Paterson, however, is remembered on a memorial at the entrance of the Panama Canal. The huge engineering feat, built some 200 years later, provides a shipping route between the Pacific and Atlantic oceans, and is proof that the trading link Paterson imagined was, indeed, achievable.

27

THE LOST US STATE OF TRANSYLVANIA

When you mention the word Transylvania it tends to send a shiver down the spine, conjuring up images of dark deeds and gory goings on. Of course, most people know that Dracula's legendary home is a real region in today's Romania. But there was once another Transylvania, and its mysterious story is almost as good a yarn as Bram Stoker's blood sucking thriller.

The story of the American Transylvania – which could easily have become one of the states we know today – begins in the eighteenth century, just months before the separate colonies broke away from Britain. It was the brainchild of a man called Richard Henderson, a judge from North Carolina who aimed to make Transylvania a new colony in its own right, the fourteenth, located where today's Kentucky now sits on the map. Though the European region called Transylvania had existed for centuries, when Henderson set out on his quest no one had even heard of Dracula. Bram Stoker's novel didn't hit the shelves until 1897.

So why Transylvania? Sylvania is Latin for a forested area. Transylvania basically meant 'on the other sides of the woods'. There was already a colony called Pennsylvania – named after its founder William Penn. The territory that Henderson was interested in lay on the other side of the

Appalachian mountains, which run north to south down the length of North America's eastern side. Hence Transylvania.

In summer 1774, just two years before the American Declaration of Independence, Henderson formed what would become the Transylvania Company, an audacious and legally dubious scheme to grab largely wild and unsettled lands to the west and form them into a new state. In March 1775, he managed to buy 81,000 square kilometres from the local Cherokee Indians for an estimated £10,000. He'd also recruited one of America's most famous sons, the explorer Daniel Boone, to help secure the territory. Boone had already attempted to open up new lands in the region and fought local Indians in a short conflict known as Dunmore's War, in which the Shawnee Indians had been forced to give up their claim to the modern region of Kentucky. Now he was asked to blaze a trail into the territory known as the Wilderness Road, along with thirty fellow settlers on what has become a famous journey through the rough terrain. Along the way, three of the party were lost to Indian attacks.

Inauspiciously, on April Fool's Day 1775, Boone reached a spot on the Kentucky River where he started building a fort and settlement called Boonesborough. Just days later, Henderson trekked through the forests to the settlement himself. Though Boonesborough still numbered just 100 people with a few outlying settlements, Henderson decided to hold a constitutional convention to draw up the framework of a government. And, on 23 May, he managed to get delegates from the wanna-be state to meet under an enormous elm tree. Amazingly, after three days of debate, an agreement was drawn up which provided for elections, balancing branches of government and even courts; a system not too dissimilar to the US constitution, which, incidentally, was not adopted until 1787.

Without wasting any time, Henderson and the rest of the Transylvania Company's owners then petitioned the Continental Congress to become the fourteenth colony, stating that they were 'engaged in the same great cause of liberty'. John Adams, who in 1776 would be a founding father of the

An anonymous sketch showing the Transylvania Constitutional Convention at Boonesborough, 23 May 1775. The abortive US state is now part of modern day Kentucky.

independent United States, warmed to the Transylvania investors, saying that they were 'charged with republican notions – and utopian schemes'. But George Washington, the first US president said: 'There is something in the affair which I neither understand, nor like, and wish I may not have cause to dislike it worse as the Mistery *[sic]* unfolds.'

There were other big problems. The colonies of both Virginia and North Carolina already had claims on the territories, and many of the settlers decided they weren't keen on Henderson's methods. He and his Transylvanian cronies were lambasted as blood suckers who merely wanted the land for their own economic gain. Henderson was soon forced to back down. In 1778 the Virginian Assembly declared the Transylvania claim void for good. Henderson himself didn't come off too badly, being awarded 200,000 acres as compensation for his efforts to colonise the West. His proposed capital of Boonesborough, however, was frequently attacked by Indians over the coming years while Boone himself was left almost penniless thanks to a robbery and some failed business ventures.

Henderson's ambitious attempts at state building are no means unique in American history. There are plenty of other bizarre states that never made it on to the familiar list of fifty that we know today. In 1850 a mining town in California called Rough and Ready tried to avoid taxes by seceding from the US. In the early part of the twentieth century residents of Oklahoma and Texas, mad at the lack of roads on which to drive Henry Ford's new cars, tried to set up Texlahoma. A similar bid in the 1970s was made by Forgottonia, a region in western Illinois which also believed it was missing out on services and development. Eventually all these attempts went the same way as Transylvania.

All is not lost for Dracula fans though. The idea of the American Transylvania lived on. Kentucky's Transylvania University, where budding vampires would surely want to study, was founded in 1780, named after the failed state. And, in 1861, a county in the state of North Carolina got the name – as a tribute to the original company.

28

THE FRENCH REPUBLICAN CALENDAR

When the French people took control of their destiny and dispensed with the services, and indeed the heads, of their ruling classes following the 1789 revolution, much was done to erase the signs and symbols of the past. In came a new system of government. In came a new way of measuring distances and weights. And, to the puzzlement of the masses, in came new days of the week, months of the year, and, most baffling of all, the way of measuring time itself. For in the new French Republic, history was consigned to the history books. 1789 became year one: the start of a brand new calendar based on a much simplified, logical metric system. This was decimalised time.

One hundred seconds made a new minute, 100 minutes made an hour, and ten hours made a day. Each new French week consisted of ten days, the ten-day cycle becoming known as a *décade*, and three *décades* made a month. The rule of ten then descended into slight chaos as twelve thirty-day months formed the year, but at least this unified months into regular periods and ended the nonsense of some months consisting of thirty-one days, others thirty, and one with twenty-eight or twenty-nine. Despite the logic with which mathematician Gilbert Romme, the main architect of the French Republican

Calendar, had designed the system – having based it on the experience on old Egyptian and Athenian calendars – precedent wasn't on his side. Since time began, calendars had traditionally been linked with astronomy. But the earth, moon and sun were no respecters of digital dogma, leading to the new calendar being five and a bit days light of the standard solar year – the 365 days, 5 hours, 48 minutes and 56 seconds it takes for the earth to circle the sun.

If the new republican regime could have altered the spinning of the earth on its axis, no doubt they would have tried. Instead, Romme and his republican committee colleagues who had been set the task of redesigning time, had to compromise. With regret, five extra days were added to the end of every year. Worse, leap years were to remain; this time tagged onto the end of the year, rather than in February (which itself now spanned two new months called wind and rain). New leap years were scheduled every four years from the end of year three (so three, seven, eleven and so on), a further deliberate attempt to break with pre-Revolutionary tradition.

In line with the new age of reason, culture, ideology and science, poet Philippe François Nazaire Fabre, known as Fabre d'Eglantine, was chosen to select names for the days and months. He chose to honour fruit, vegetables, animals and even pieces of agricultural equipment for every one of the 366 days. The old calendar's system of holidays, inextricably linked with Catholic saints' days, even Christmas, were replaced with a series of days at the end of the year thanks to the five lunar days that Romme couldn't avoid. These celebrated different liberties, including one that allowed Frenchmen to libel one another without fear of legal retribution – quite something when someone could be condemned to the guillotine for anti-revolutionary sympathies.

Traditional Calendar	Republican Calendar	Meaning
September / October	vendémiaire	Wine harvest
October / November	brumaire	Mist
November / December	frimaire	Frost
December / January	nivôse	Snow
January / February	pluviôse	Rain
February / March	ventôse	Wind
March / April	germinal	Germination
April / May	floréal	Flowering
May / June	prairial	Meadow
June / July	messidor	Harvest
July / August	thermidor	Heat
August / September	fructidor	Fruit

The five new holidays at the end of the year, beginning the day after 30 fructidor:

1 Fete de la vertu (holiday of virtue)
2 Fete du genie (holiday of genius)
3 Fete du travail (holiday of work)
4 Fete de l'opinion (holiday of opinion and free speech)
5 Fete des recompenses (holiday of rewards)
6 Jour de la revolution (day of the revolution) (the leap day)

With numbers and names decided, all that was now needed was to select a start date, and here too anxious debate raged. Some revolutionaries believed that it should coincide with the overthrow of the monarchy – the Revolutionary Calendar – others that it would be more fitting for it to commemorate the new republic, which happily coincided with the autumn equinox, on 22 September 1792. The second option won the day and the calendar was formally adopted on 24 October 1793 (backdated to 22 September the previous year), although because getting the message around the country from Paris took a little time in the eighteenth century, different parts of the country began to use the calendar at different times.

The planet's elliptical journey round the sun caused a further complication. Under the Gregorian calendar that had been used until then and which we use today, the equinox can fall on either 22 or 23 September. Clearly, for a logical system, this is utter lunacy and on this matter Romme was not going to give in to physics. He reasoned that once the Republican Calendar was introduced, provided it was ruthlessly enforced, 22 and 23 September would be an irrelevance. Henceforth the new date of 1 harvest would synchronise the start of the New Year *and* the equinox.

The period from what is now 22 September 1792 to 21 September 1793 became year one (expressed as roman numeral I), presenting a further complication; one not experienced for 1792 years. History had to be re-dated. Calculations were made to show that the Creation occurred 5,800 in advance of the Republic, Noah built his ark 4,144 years previously, and printing had been invented 330 years before. In year VIII of the new era, an almanac of practical information for the household – a common publication in French districts of the period – looked to the future too, providing a ready reckoner that helped people calculate the average number of years they had left to live at any particular age.

With the calendar up and running, the authorities waited for the people to take it to their hearts. But it only caused trouble, confusion and frustration. Decimal weeks now meant that weekends fell every ten days, instead of every seven. Weekly markets, events, meetings, even paydays suddenly became less frequent. And hours almost doubled in length, which made working days disagreeably long for the French. With all these new names, many people quite literally didn't know what day of the week it was, and the still largely Catholic population found the abolition of Sundays particularly hard to fathom. Pleumartin, a village in Poitou-Charentes where couples traditionally married on Wednesdays, discovered that the newly assigned day Quintiti turned up only every tenth day, putting a pressure on slots at the register office.

Discontent bred rebellion, but attempts to circumvent the calendar were ruthlessly repressed. Police swooped to stop markets being held on the wrong days or at the wrong times, and even to enforce the new day of rest, the décadi. Brave traders and shoppers sometimes fought back, attacking gendarmes, but this was risky. At the height of the French Terror, anything that could be interpreted as anti-revolutionary could be severely punished. With wonderful irony, Romme himself eventually fell victim to the Terror. At the time of the revolution he had voted for the king's execution, but since then he had made protests against the new regime. Arrested, tried and sentenced to death, he committed suicide on his way to the guillotine in 1796, using a small knife he had secreted about him rather than wait for the big chop. At this, he became the Martyr of Prairial – prairial being the ninth month of the Republican Calender. .

The calendar didn't survive much longer than its chief architect. Sensing its unpopularity, Napoleon later agreed an accord with the Pope, and at midnight on 31 December 1805 (10 Nivôse, year XIV), time was called on the French Republican Calendar. After nearly thirteen years, the country aligned its clocks with the rest of the world on 1 January 1806. This wasn't quite the end. The calendar enjoyed a very brief revival later in the century when the Paris Commune government, which briefly ruled the capital in 1871, restored it for ten weeks between 18 March and 28 May.

29

LATIN MONETARY UNION

If nineteenth-century history is anything to go by, changing a workable system that is universally understood and with which most people appear happy was becoming a French characteristic. For what they had already done successfully for measures, but disastrously for dates, French republicans now attempted with currency. In short, out with the old, in with the new. Gold – the international indicator of national wealth and the backbone of banking – would no longer be permitted in French coins from 1803.

To 'banish gold', as one historian called it, was thought by many economists to be madness, especially at the start when the Napoleonic wars needed to be paid for. But once the wars ended and the French economy returned swiftly to its feet, neighbouring countries started to look on with increasing admiration. Starting to base their own economies on the franc, a currency consolidation began that would last throughout the middle 1800s. Soon a formal monetary union was underway, and France was at its heart.

This initiative, called the Latin Monetary Union (LMU), had its origins in Europe's changing political geography and the quest for economic stability across the continent. Within two years of Belgium separating from Holland in 1830, it had ditched the Dutch guilder for francs. Italy, going through a seemingly never-ending series of revolutions on its way to unification, aligned with the franc in 1861. With other countries happy to link to a single currency too, a path was set for a more formal union, with Spain, Greece, Romania, Bulgaria, Venezuela, Serbia and the Vatican going Francophile a year after its introduction in 1866. The currency union staggered on for fifty years, but was manipulated by a misjudgement about the value of gold relative to silver years before it even began.

Back in 1803 France decided that all newly minted coins, up until then a mix of gold and silver, should be made out of silver only. At this time, coins in many countries, including France, contained real gold and silver in carefully regulated amounts, based both on the weight of the precious metal and its fineness. And because under a system called 'free coinage' anyone could have their own gold and silver pressed into coins, it was crucial that the value between the two metals was correctly established. If there wasn't 'parity' between the two, one could be used to buy the other very cheaply which could be then sold elsewhere for a profit. This interchange between silver and gold was known as bimetallism, and because the parity between the two metals was wrong at the start, the Latin Monetary Union was arguably doomed from inception.

LMU experts Kee-Hong Bae and Warren Bailey explain:

> The weight, fineness, and denomination of coins defined a 'mint ratio' for the value of gold versus silver. If, for example, a silver one-franc coin contained five grams of silver and a gold twenty-franc coin contained six and two-thirds grams of gold, the implied mint parity was 15 to 1.

If this was complex, it didn't matter. All people really cared about was whether their money kept its value. Nevertheless, 15:1 was a crucial ratio, but it wasn't the one that the French set when they changed their currency in 1803. On 7 germinal XI in the French Republican Calendar, the country started producing new francs containing 5 grams of silver 0.900 fine;

a bimetallic ratio that slightly overvalued silver at 15.5:1. This was a mistake that was to prove costly further down the line. Furthermore, although France stopped producing most gold coins, old ones remained legal tender and a new coin for high value transactions, named the 'germinal franc' after the month of germinal in the Republican Calendar, was pressed to mark the occasion.

As time went by and other countries joined the francozone, each nation decided on the fineness of silver in its own coins. In Italy, 0.835 of silver to alloy was agreed, while Switzerland chose 0.800. As monetary union took a step closer, every country wanted its own fineness to become the adopted standard to save it having to mint new coins. France, most of all, was determined to get its way and reduced the fineness of the silver in its coins to 0.835 in 1864, just before the introduction of the union. Other countries followed suit. Eight coins were issued throughout the four founder nations of France, Belgium, Italy and Switzerland, when the Latin Monetary Union was formalised in August 1866.

The United Kingdom, true to form, resisted, saying while it would like not to be excluded, the LMU was a continental idea and the UK would prefer instead to go it alone and concentrate on strengthening the pound. Changing its mind within two years, the chancellor of the exchequer, Robert Lowe, tried to agree a deal with France to join, but was rebuffed. Even so, some key figures in the UK were being turned onto the idea as a logical extension to decimalisation that was being debated in parliament at the time. A senior banker and an advisor to the ruling Liberal government, Lord Overston, proposed that the pound be equalised to 25 francs. A new 20 franc coin should be introduced too, to be known as a 'Queen' to replace the sovereign he said. Overston's idea was laughed out of parliament, though he persisted in his case at home and abroad.

Although the UK was on the sidelines, eighteen countries eventually joined the LMU. But the rules were complicated and inconsistent, and there were too many loopholes. For economic reasons, a cap was placed on the amount of money each nation could produce: no more than six franc coins for each person in their country. And each member could strike low-value silver coins containing less silver than their face value for use in their country alone. Under LMU regulations, government bodies and local authorities were obliged to accept silver coins from any country in the monetary union up to a value of 100 francs per transaction. With differing amounts of silver in the coins, people flocked to change low-ratio silver coins to ones with higher silver content, causing silver to become overvalued. Whole countries took advantage. Coins from nations with low silver content were exchanged in other countries for gold. Germany, not part of the union, found a rich seam in sending officials to Paris with silver ingots where they would be minted into five franc coins, which were exchanged first for banknotes, then for gold. Even the Pope was at it. The papal state's treasurer, Giacomo Cardinal Antonelli, ordered coins to be minted that were light on silver and then exchanged them for coins from other countries that had the correct amount of the metal.

France soon experienced a run on its iconic Napoleon 22-carat gold coins. At the time, a gold Napoleon had a face value of 20 francs. People found they could mint four 5-franc coins with 16 francs worth of silver, buy a Napoleon and make an immediate profit. During the five years between 1853 and 1857, France lost 1,125 million francs of silver.

The manipulation of the exchange rate between gold and silver continued unabated until 1874, when the union suspended the convertibility of gold to silver and stuck to the gold standard, the international monetary system based on fixed weights of gold that the UK had adopted in 1816. For all the countries in the LMU, it was a humiliating climb down, if not quite the end for the LMU. After four years, the temporary suspension became permanent – and although no new silver coins were struck in the LMU, the old ones continued to be legal tender. The widening gap between the value of gold and silver and the impact of the Great War made survival of a currency union almost impossible. Although the LMU's formal demise did not come for a further fifty years – until 1926 – in practice, silver-light francs exchangeable freely across eighteen nations had ceased to exist long before.

30

A TAX ON LIGHT AND AIR

Coins that were light on silver or gold were not just a problem for the French. Through the ages, entrepreneurs with an eye for a quick buck have made money for nothing, shaving small fractions off coins to smelt into new currency. 'Clipped coins' introduced new cash into economies, but as well as being illegal, they caused inflation, leaving people feeling poorer. So when, in 1686, England faced a currency crisis and was carrying a huge national debt, the king, William III, vowed to solve the coinage scandal once and for all. As well as minting new coins, he would plug the hole in the country's finances with a fair way of raising new revenues – a tax on light and air.

England was in dire need of money. The king's two-pronged tax-and-mint strategy would restore confidence in an economy burdened with so many clipped coins no one really knew how much silver they had in their pocket. But issuing new currency would be costly. Coins had to contain the correct amount of metal, which meant buying silver and incurring more debt – and cash was something William didn't have. Even if he did, he wasn't sure that it was worth the value stamped on it.

If minting new coins to solve the clipped coin crisis was relatively simple, how to tax a recalcitrant nation was less so. A widely despised hearth tax, in force for a little over twenty years, had only recently been repealed. If an Englishman's home is his castle, people said that tax assessors – who were known as 'chimney men' for this purpose – had no right trooping in to tot up hearths, stoves and chimneys. Older taxes and foreign levies had their attractions. A former tax on bachelors, for instance, promoted marriage and procreation when more people were needed to fight wars, but the king would wait till 1695 when skirmishes with the French were hotting-up again before inflicting that one on England. A beard tax carried a certain perception of even-handedness, as Henry VIII, who introduced it, wore one himself, but this was not on the cards for William. Increasing the tax on playing cards would raise insufficient revenues; whilst income tax was completely out of the question. When William considered it, the collective nation almost choked on its breakfast. The very idea that the government could pry into what people earned was an outlandish invasion of privacy. So William looked out of his palace and hatched a plan. A window tax, though as intrusive as the hearth tax before it, would be accepted with marginally less opprobrium, he thought.

Casting aside the usual complaints – it would cost too much to collect; it was unnecessarily bureaucratic; there were too many loopholes – William's 'Duty on Lights and Windows' became law in 1696. All houses, except those of the very poorest, paid 2s a year, regardless of the number of windows. Dwellings with between ten and twenty windows paid 6s, and anyone with more than twenty windows paid a total of 10s. If this seemed clear enough, the incomprehensibly complex rules were eminently breakable.

Tax evasion began almost at once. Homeowners took to bricking up or painting over windows. Some rooms were exempt, leading people to dedicate the most improbable of spaces as pantries or grain stores. As assessments were infrequent, windows could be bricked up one day and un-bricked the next. Bribery was commonplace, especially as the local JP, usually a well-connected man of property known to the wealthier landowners, acted as the assessor and would often

The window tax, introduced into England in 1696, meant that many owners bricked up existing apertures.

turn a blind eye to the glazing extravagances of influential neighbours. Even the definition of what was or was not a window was a troubling matter of conjecture that could go either in the landowner's favour or the taxman's. According to contemporary dictionaries, the word derived from 'wind-door', signifying 'any aperture in a building by which light

and air are intro-mitted'. Tax inspectors adored this broad
definition as it covered just about every hole in the wall. If
they didn't like the homeowner much, it could be costly, as
one unfortunate gentleman with a severe damp problem discovered. On the advice of a sanitary expert, a Mr Williams
installed four perforated zinc plates to improve ventilation in
his home, only for the taxman to assess it as a new window and
place him in a higher tax bracket. When he appealed, magis-
trates decided that both parties were wrong. Four zinc panels
created four new windows, they said, and they increased Mr
Williams' tax further still.

Such cases were widespread until a revised law clarified the
position in favour, quite naturally, of the taxman. Individual
panes cast in one frame were classed as separate windows, even
if they didn't have glass. And even a hole could be a window.
In one case, a space in a wall used for shovelling coal was taxed.
The taxman nearly always wins.

But not quite. The inventiveness of taxpayers to thwart the
authorities took such hold that by the 1720s revenues were in
decline. A treasury review in 1729 found 'a sorry story of windows stopped up, of JPs obstructing the work of assessment, of
surveyors lazy and incompetent'. The law was tightened again,
with stiffer penalties for anyone caught blocking and reopen-
ing windows. The tax now covered:

> skylights, windows, or lights, however constructed, in stair-
> cases, garrets, cellars, passages, and all other parts of the
> house, to what use or purposed soever, and every window
> or light in any kitchen, scullery, buttery, pantry, larder, wash-
> house, laundry, bakehouse, brewhouse and lodging room
> occupied with the house, whether contiguous to or dis-
> joined from the dwelling-house.

Construction of outhouses, an unintended consequence of
window tax and from which they were exempt, slowed.

With the country needing more money again, revenues
started to rise. William III was long gone, but by this time the

Seven Years War in Europe needed funding and other taxes like the cider levy and one that funded the war in America had been repealed. In 1766, window tax was extended again, bringing in homes with just seven windows. The result changed housing and architecture in Britain at all points on the social scale. Although window tax had originally been designed as a progressive tax that hit the rich more than the poor, because windows in whole buildings were assessed, not those in individual garrets or flats, landlords of tenement houses found themselves heavily taxed. To compensate, they upped rents on existing properties or boarded windows up, leaving tenants now not just in squalor, but in dark, airless squalor to boot. In Edinburgh, entire rows of houses were built without a single window in the bedrooms. Elsewhere, bedroom windows were blocked or painted on the logical assumption that the hours spent there were dark anyway. By contrast, at the top end of the scale, the landed gentry found a new way of demonstrating their wealth. Designers of opulent, ostentatious country homes, palaces and avenues packed in as many windows as they could. The master would willingly pay the tax as long as the neighbours noticed.

Still the government wanted more. In the 1780s, with the French sabre-rattling and Britain on a war footing, taxes rose again. Fortunately for the government, the prime minister, William Pitt the Younger, had a talent for taxes. A guinea imposed on the white powder gentlemen used to give their wigs an off-white hue became the wig tax, but raised an insignificant amount and upset the gentry. Hat tax similarly targeted the rich who, if they wanted to get ahead, got a hat – or rather several hats, one for every occasion. Tax-evading hat wearers and milliners faced the threat of execution. Each piece of men's headwear had to have the Revenue's stamp on the inside. Forging a stamp, whether on hats, wig-powder packaging, playing cards, or indeed on anything taxable, could, and sometimes did, lead to the gallows.

Pitt's initiatives didn't stop with headwear. A tax on sporting dogs came in 1786. A brick tax at 5s per 1,000 bricks raised yet more money (but also caused manufacturers to make buildings from larger bricks). A levy on 'watches and persons in possession of clocks' taxed time. Even soap, a luxury in the Georgian era, came with a contribution to government coffers. But the most galling tax of all was income tax – the very charge that even William III couldn't bring himself to introduce all those years ago. Income tax caused untold consternation among the landed voting classes and, when introduced in 1799, didn't last long. In 1815, to whoops of joy from MPs – the very people paying most – it was repealed and all documentation relating to it pulped. Income tax wouldn't be seen again until 1842 and even then, like now, it was a 'temporary tax' that would only last a year (but is renewed at each Budget).

With income tax gone, window tax rates again rose to cover the deficit. But it was despised all the more and tax inspectors increasingly sought police protection when assessing homes. Even they began to despise it. At one end, they were taxing their important neighbours; at the other they were contributing to increasingly evident health hazards caused by lack of light and air. In greater numbers, working people lived in damp, unsanitary conditions with inadequate ventilation. Much of the trouble was ascribed to window tax. Amid worsening conditions, respected architect Henry Roberts reported on the link between window tax and poor health. His findings were scathing; the link indisputable. In the new Victorian age with the world about to descend on the 1851 Great Exhibition, this was a time to clean up. Just before the exhibition opened, window tax was repealed, although a tax on glass remained for six more years. New ways other than window tax would have to be found to fund government. And they have yet to run out of ideas.

31

PAXTON'S ORBITAL SHOPPING MALL

No one could call Joseph Paxton a failure. He was the man who designed the Crystal Palace – that glorious glass hall which hosted the Great Exhibition in 1851. Paxton's magnificent structure, erected in London's Hyde Park, was visited by six million people during the event. It was certainly much more elegant than most of the other designs which had been on the table. They included an enormous brick building with an iron dome, suggested by none other than Isambard Kingdom Brunel. In fact all of the 245 designs which had been received for the proposed exhibition hall following an international competition were deemed so impractical that when Paxton published his own sketches for a pre-fabricated glasshouse in the *Illustrated London News* at the last minute, his scheme was snapped up.

The hall, later dubbed the Crystal Palace by the satirical magazine *Punch*, was nearly 2,000ft long, 408ft wide, 108ft high and made out of 4,500 tons of iron along with nearly 300,000 sheets of glass. It was put up in just eight months from August 1850. Amazingly, it was only ever meant to be temporary. When the exhibition came to an end no one quite knew what to do with it. Paxton himself mooted that it stay where it was and be turned into a steamy winter garden. One architect even suggested rearranging the panes of glass to turn it into a huge 1,000ft glass tower.

Eventually, the palace was moved, pane by pane, to Sydenham Hill in South London where it reopened in 1854 and stood overlooking the metropolis until it tragically went up in flames in 1936. Not content at wowing the world with his Crystal Palace, Paxton had another, even grander idea up his sleeve. By the mid-1850s Paxton, who had started out designing gardens, was riding high on the success of the Great Exhibition. He'd been knighted for his efforts and had become an MP representing Coventry. In 1855 he felt bold enough to present to his parliamentary peers a plan for the Great Victorian Way, a project which would, he said, 'make London the grandest city in the world' – and make the Crystal Palace seem trifling. The Great Victorian Way was, in effect, a vast, enclosed, shopping mall combined with a ring road and circular railway system. This 'Grand Girdle Railway and Boulevard Under Glass' would encircle London, above ground, in a 10-mile loop crossing the Thames and back again to join up each of the main railway termini. It would also have a mile-long branch line using a third bridge to take travellers to the Houses of Parliament and Victoria Street.

There would be an elevated glazed central walkway-cum-roadway in the middle for pedestrians, wagons and coaches. A breathtaking eight railway lines would be carried in closed galleries around the outside, catering for both express and local stopping services. The trains were to be propelled by atmospheric pressure, a new, somewhat faltering, technique. Paxton said the engineer Robert Stephenson, son of railway pioneer George, had assured him it would work. These new lines would drastically cut travelling times across the clogged city (which did not yet have an underground railway). On the Great Victorian Way you would be able to travel from Bank to Charing Cross in just five to six minutes reckoned Paxton, a feat that can hardly be achieved even today.

Within the arcaded central avenue of the great girdle there would be space for shops, hotels, restaurants and even houses.

Paxton saw it as a kind of modern version of the medieval London Bridge which, in its time, once had houses and shops along it. All 10 miles of his structure would be 72ft wide and 108ft high and, like the Crystal Palace, would be built from mainly iron and glass, with a great glazed roof all the way round. Paxton's love of glass was fuelled both by his early work on large garden conservatories and by the fact that its price had fallen, making it cheaper to use in large building projects.

When he appeared in front of a select committee in the summer of 1855 to be quizzed on the proposal, Paxton was keen to point out that no 'important street' or 'valuable property' would be knocked down and that his great girdle wouldn't obstruct any existing major thoroughfare. He even claimed that the residences that were to be embedded in the structure would 'prevent many infirm persons from being obliged to go into foreign countries in the winter'. In an article that year the *Civil Engineer and Architect Journal* refers to the Great Victorian Way as The Paxton Arcade and celebrates its 'brightness of light and immunity from weather as a promenade, and drive, in summer and winter'.

The cost was eye-watering though. His Crystal Palace had cost only £79,800 to build. The Great Victorian Way, admitted Paxton, would need an outlay of £34 million. In twenty-first-century terms, with inflation taken into account, you're looking at a project costing billions. At this price Paxton knew it would need not only the blessing of government but its financial backing too, though he proposed that a private company actually carry out the work.

At first his fellow MPs seemed favourable and certainly gave a lot of consideration to the plan. Something like it was needed. Congestion was already a thorny problem in a city which had swelled in the preceding few decades thanks to industrialisation. As Paxton pointed out, it sometimes took longer to cross central London than to travel from the capital to Brighton. And Paxton's plan had been well thought through. He'd worked on ventilation systems, drawn up plans for the route's bridges and even chosen the kinds of tiles he was going to use for decoration. Not everyone thought the Great Victorian Way was a good idea though. The new *Daily Telegraph* newspaper, born in June 1855, stated in a July article that: 'We must protest against being disturbed in our slumbers by a whistle and a roar overhead.'

However favourable some found the plan, the government were ultimately unwilling to take on such a huge financial burden, even for the tried and tested Paxton. Just a few years later the 'roar overhead' began to be replaced by a roar underneath, as the first of the London Underground lines was built, following the same idea of connecting the big railway hubs that Paxton had. The great man lived just long enough to see this much less dainty version of his vision become reality when The Metropolitan Line opened in 1863.

32

CINCINNATI'S SUBWAY TO NOWHERE

There are long platforms with huge flights of steps and miles of tunnels. But no trains ever arrive, there are no passengers to get on them and it's been that way for more than eighty years.

Many of the world's major conurbations have some form of underground railway, metro or subway. So does the city of Cincinnati in the US state of Ohio, with one big difference. Its subway system, much of it still intact beneath the streets of the modern city, has never been used. As white elephants go, this is a monster. Far from being one of crowning glories of the Queen City, as Cincinnati is known, the subway has since been described as, 'one of the city's biggest embarrassments'.

The tale of this ghost network goes back to 1884 when the local *Graphic* newspaper suggested draining a fetid canal that ran through the metropolis and replacing it with a subway. It wasn't until the early 1900s that a plan to make the subway a reality was produced and citizens of Cincinnati finally got their say on the idea in 1916. By then cities such as New York, and of course London, had underground or subway systems with some sections running above ground, others below. Cincinnati, the tenth largest city in the USA, was expanding fast and aimed to join the big players. Despite 200 miles of tram tracks, it desperately needed to solve its traffic problems too. So residents went to the polls and overwhelmingly approved the construction of a 16-mile subway which would loop round the centre of the city and its suburbs, at a cost of $6 million paid through an issue of bonds. A basic fare was even agreed – 5 cents a journey.

Then, crucially, in April 1917, America entered the First World War and work on Cincinnati's subway was put on hold. Once the war was over, local politicians decided to continue with their grand plan, even though inflation saw costs of steel and concrete soaring well above the plan's original estimates. On 28 January 1920 excavation began. Completion of the project was expected in five years. Financial pressure soon meant that the original plan was reduced to 11 miles. But by 1923 the underground section of the subway had been finished. Costs kept going up. Yet so did the building work, thanks to the municipal authorities who still had a pot of cash and lucrative contracts to hand out. The city, which had been built on pork packing, was succumbing to pork barrel politics. By 1924 they were up to nearly 7 miles.

Then, in 1926, the new no-nonsense mayor, Murray Seasongood, revealed the bad news: another $10 million would be needed to save the subway. There was also a much more important player on the scene – the automobile. When the subway had first been touted there were few cars on the streets. But the decade of the 1920s saw the burgeoning of the American love affair with the car. Ownership of cars in the Cincinnati region went up by 126 per cent between 1921 and 1926, making the subway seem less essential.

In 1927 work on the subway was brought to a halt. In 1928, the airy boulevard above, the Central Parkway, was opened to much celebration. Most of the revellers would soon forget that below their feet lay a near-complete subway network on which the city had blown its cash. What remained were miles of empty parallel tunnels, with three complete stations situated in the section that ran underground at Liberty Street, Race Street and Brighton's Corner, all now concealed beneath the hustle and bustle of the city's roads. There were also three sta-

The US city of Cincinnati's subway under construction in the 1920s. The tunnels would never be used.

tions above ground ready for action at Marshall Street, Ludlow Avenue, and Clifton Avenue, which were later demolished, to make way for more roads. Once construction had stopped hopes of getting the project back on its feet were totally dashed following the 1929 Wall Street Crash and subsequent Great Depression. A line was finally drawn under the project in 1948.

In the decades that followed, debate has raged about what to do with the empty subway system. Proposals to use the tunnels as roads or for new rail systems faltered. During the Cold War a part of the system was even turned into a nuclear fallout shelter, also since abandoned. A massive wine cellar was once mooted and the atmospheric tunnels were reportedly even considered as sets for the Hollywood blockbuster Batman Forever, but even the movie moguls pulled the plug.

So there they still sit, several miles of defunct tunnels and stations minus the steel rails and trains that will never come; a monument to the financial and political perils of town planning. It took until 1966 for Cincinnati to pay off the original bonds for the subway at an extra cost of around $7 million.

The administration has since shelled out more cash to keep the subway mothballed, partly to make sure the road above doesn't fall in. The alternative is spending huge sums to fill in the tunnels.

Cincinnati might be the best example of an underground system which was built but never used. But other networks have their ghosts too, including London's underground. If you look carefully in the suburbs you can still see the remnants of the extension to the Northern Line that was partially built in the 1930s. The Northern Heights scheme would have seen the line extended to Bushey Heath with stations at Elstree South and Brockley Hill. A route was laid out and bridges built. You can even see some of the arches put up to carry trains into the Bushey Heath Station that never was. But once again a war intervened – the Second World War. Afterwards, new green belt planning restrictions meant that the area wasn't going to develop as had once been expected, so there just wouldn't be enough passengers for the line.

Around the world there are more ghostly stations in subway systems. Paris has a station in its network called Haxo, completed but never used as it was eventually deemed unprofitable. Stockholm has its own example, Kymlinge. Yet none of these has ever been the millstone that the Cincinnati subway became for city officials. It's hardly surprising that tales emerged of the tunnels being haunted by the spirits of construction workers who were unlucky enough to die during the building of their transit system. You could forgive them for feeling a little disgruntled. After all, its demise wasn't their fault. As one Cincinnati council member summed up the subway in 2007: 'It didn't go anywhere, but it was built well.'

33

IS IT A TRAIN OR A PLANE?

In the 1920s railways still ruled when it came to mass transport over big distances. And, with some rare exceptions, steam power still ruled the world's railways. Yet inventors like George Bennie were looking for new ideas to revolutionise rail travel. Some were experimenting with diesel designs, others with electric units. Bennie, the son of a Glaswegian hydraulics engineer, had an idea to combine the sleek lines and technology employed by the age's new aircraft with what he saw as the reliability of a land-based transport system. And so, in 1921, the Bennie Railplane was born.

This was a far cry from some of Bennie's other inventions. His improved golf putter for example, which featured a special boss on the face of the club for extra accuracy, had not proved eye-catching. His futuristic, streamlined Railplane, on the other hand, provided a startling spectacle. And, on 8 July 1930 when the gleaming prototype was shown off to the press and potential investors, Bennie boasted that his Railplane would be capable of 120 mph. This at a time when express steam trains were only averaging 80 mph.

The Railplane looked a bit like the kind of monorail used in some cities today. But the cars were driven by four bladed propellers at either end powered by either live electric rails or combustion engines. These cars were suspended from a large metal gantry with a guide rail at the bottom. Bennie intended that hundreds of miles of this gantry would be constructed across the country above existing railways. This would free the traditional railways up for freight, easing congestion. Meanwhile passengers would be hurtled to their destinations at lightning speeds in modern comfort high above the tracks.

Bennie claimed the Railplane would be economical to build too. He estimated the construction cost of an ordinary railway at £47,500 a mile, while his Railplane would cost just £19,000 a mile. Most importantly it would be almost as fast as going by aircraft. At the time these were pretty ponderous, had limited capacity and were often grounded in bad weather. Bennie believed his Railplane would be safer than taking to the skies yet be able to whisk travellers from, say, Glasgow to Edinburgh in just twenty minutes and from Glasgow to London in three and a half hours.

After years of finessing the design, with the help of an engineer called Hugh Fraser, Bennie persuaded one of the biggest rail companies in the land, the London and North Eastern Railway, to let his company construct a test line over a rail siding at Milngavie in East Dunbartonshire. The 426ft line was 80ft across and 16ft from the ground. His prototype car was 52ft long, 8ft in diameter, weighed in at 6 tons and was designed to take fifty passengers. The look of the thing owed a lot to the airships of the time and indeed its engineers, William Beardmore and Co, had built the R34 airship, which made the first airborne, east to west, Atlantic crossing in 1919.

Newsreels of the Railplane's launch show plush interiors with comfortable armchairs and a saloon. The sliding doors even featured elaborate stained glass. Technologically the Railplane was well thought out. As a train approached the station the front propeller would stop while the back one was switched on to achieve reverse thrust, with brake shoes gripping the rails. But one of the most positive aspects would be the lightweight nature of the whole design that meant there was little track resistance. In a brochure for his Railplane,

George Bennie's Railplane prototype running on a test track in Milngavie, East Dubartonshire, Scotland. The Railplane was a fusion of aircraft and railway technology.

which harks back to the glorious rail breakthroughs by the likes of George Stephenson, Bennie also points out that the Railplane can tackle hills easily.

Things looked positive. One reporter described travelling on the Railplane in a test run as a 'sheer delight' and luminaries such as prime minister Ramsay MacDonald took an interest. Council officers in Blackpool, Lancashire were keen to investigate how the Railplane could be used for a link to nearby Southport, and there was even genuine interest from Palestine.

Sadly, like many good ideas, the timing was off. The Great Depression meant money was tight and, in the end, no orders came in quickly. Perhaps Bennie was too ambitious. He made a submission to the government in 1935 to build a line from London to Paris involving a seaplane link across the channel. Bennie reckoned that his link would speed people between the capitals in 140 minutes. A contemporary aircraft would take 225 minutes. This proposal hit the bureaucratic buffers too. So convinced had he been in success that Bennie had already invested £150,000 of his own money in the project.

By 1937 he was declared bankrupt and with the Second World War on the horizon, revolutionary schemes like Bennie's were destined to end up sidelined.

There had been other attempts to fuse aircraft and rail technology abroad, with equally unconvincing results. A prototype Russian rail plane called the Aerowagon, powered by an aircraft engine and propeller, derailed while being tested in 1921; killing everyone on board including its inventor, Valerian Abakovsky. Franz Kruckenberg had more success launching his 'rail zeppelin' in 1929; it even reached an impressive 160 mph. Technical issues, however, including the fact that the era's tracks really couldn't take a train going that fast, meant it was a commercial non-starter.

Despite attempts to re-boot the Railplane after the war – Bennie even took his ideas to the Iraqis – the world had moved on and the remains of his Railplane prototype and its test track were finally demolished for scrap in 1956. Bennie himself died a year later aged just 65, having, by some accounts, given up designing trains to become a herbalist. His design did have its flaws. It was noisy and his plan for junctions involved time consuming turntables. Doubts remain about its viability for long-distance travel, which was Bennie's real dream, and surely its unsightly gantry would have thrown up plenty of planning issues. It was probably more costly to build than Bennie had calculated. Nor did he foresee technological improvements by planes or the growing expansion of road travel.

In comparison to some other railway innovators of the time, however, Bennie's ideas look positively cautious. How about a train powered by rockets? Sounds like a one-way ticket to disaster, but Fritz von Opel, who invented it, thought otherwise. He was grandson of the Adam Opel, the founder of the famous car company, and in 1928 he tested a rocket-powered car, the Rak 1. Later the same year he decided to put his rocket car on rails. On a closed set of tracks his Rak 3 reached the amazing speed of 157 mph. Attempting to beat the record a few weeks later his new improved Rak 4 exploded into smithereens, effectively putting an end to any similar experiments.

34

HOW THE CAPE TO CAIRO
RAILWAY HIT THE BUFFERS

In 1877 when British-born Cecil Rhodes was 24 years old and studying as an undergraduate at Oxford, he made an extraordinary statement as part of his will. It read: 'I contend that we are the first race in the world and that the more of the world we inhabit the better it is for the human race.' Brought up in Southern Africa, Rhodes was an unashamed colonialist who helped the British Empire expand its territories in the southern part of the continent in the latter part of the nineteenth century. He gave his name to the colonies of Southern and Northern Rhodesia, now Zimbabwe and Zambia, and also set up the famous diamond company De Beers. You'd have thought that would be enough, but not for Rhodes. It was an unflinching belief in the superiority of British rule, an iron will and a canny entrepreneurial brain that led to Rhodes' idea to link up British territories across Africa, from the very north to its southernmost tip. The backbone of this plan would be the Cape to Cairo railway, a line more than 5,000 miles long which would have to cross some of the most remote, most disputed and most physically challenging territory anywhere in the world.

The odds against such a railway were always formidable. In Russia the Trans-Siberian Railway, of a similar distance, was

The Rhodes Colossus, an 1892 cartoon from *Punch* magazine showing empire-builder Cecil Rhodes astride Africa.

well under way by the turn of the nineteenth century. But Rhodes' railway would have to be built in a climate just as punishing and its workers would have to contend with hostile locals and dangerous animals, all against the backdrop of political turmoil as the great powers scrambled to conquer the remaining parts of the continent.

In the 1890s Rhodes certainly looked like a man that could make it happen. He had become prime minister of the British held Cape Colony in the very south and developed a serious reputation as an Empire builder by securing new territories for the British crown to the north. These became known as Rhodesia, under the auspices of a company he'd set up, the British South Africa Company. A cartoon which appeared in the satirical *Punch* magazine in 1892 was titled *The Rhodes Colossus* and portrayed him as a giant astride the whole continent.

In the same period, Rhodes furthered the construction of railways in Southern Africa and raised much of the finance himself. By 1897 a railway had got as far as Bulawayo, Rhodesia, 1,360 miles from Cape Town. The moment was celebrated with great pomp and Queen Victoria even sent a telegram of congratulations. This achievement was largely down to the efforts of a thrusting railway engineer, George Pauling, who pushed the track hundreds of miles through the inhospitable Kalahari Desert at breakneck speed. In one day alone 8 miles of track had been laid. Rhodes, suffering from fever, wasn't at the Bulawayo celebrations, but also sent a telegram vowing to forge the railway onwards to Victoria Falls.

At the other end of the continent there was another breakthrough. In September 1898, British forces under the command of Lord Kitchener secured control over Northern Sudan at the Battle of Omdurman, where he defeated the local Mahdi's forces. And the same month, a tussle with the French in what was called the Fashoda Incident secured the south of the Sudan, putting an end to French hopes of building their own railway east to west across Africa. This enabled a railway and ferry link to extend 1,000 miles south from Cairo to Khartoum in Sudan, completed in 1899. Following the victories Rhodes had even cheekily cabled Kitchener, saying: 'If you don't look sharp … I shall reach Uganda before you.' Uganda, in the middle of the continent, had recently come under British rule too. Kitchener swiftly sent a telegram back saying: 'Hurry up!'

By the time Rhodes died in 1902, aged just 49, the Boer War had seen Britain annex the independent Boer republics, leading ultimately to the Union of South Africa and the consolidation of British territory in the region. Rhodes may have gone, but Pauling carried the torch onward. In 1904 the section extending northwards with a spectacular bridge to Victoria Falls was completed, allowing tourists easy access to the natural wonder for the first time. The project was another 300 miles towards its goal and the Cape to Cairo dream still seemed very much alive. But there was a big problem ahead – the Germans. They controlled German East Africa, modern day Tanzania. While the Kaiser did allow Rhodes to build a telegraph line through his territory, he wasn't going to let the British connect up their railway lines through it.

George Pauling continued to push north nevertheless, building the line to the border with Belgian-controlled Congo instead. This came with the added temptation of opening up new mineral mines. In 1908 the *New York Times* reported that there were only 700 miles to go before it would be possible to make the trip by lake, river and railway from Cairo to Cape Town. And, in December 1917, newspapers announced that the railway had reached Bukama on the banks of the navigable Upper Congo where a steamer link could take travellers further north to Lake Tangynika. A person could now travel by train 2,600 miles north of Cape Town on the Congo Express in just six days. The defeat of Germany in the First Word War then seemingly gave Rhodes' vision an extra boost as German East Africa fell into British hands. Rhodes' idea of a continuous line of unadulterated British territory stretching from the Cape to Cairo had come true.

At the same time his railway seemed to be dropping off the priority list. By the 1930s it was theoretically possible to make the journey overland using a series of railways, ferries and buses entirely through British held territory. A few more sections of line had been built in Uganda and Kenya but a line was never pushed through between Uganda and Sudan, and few were prepared to undertake the arduous journey. It seems that much of the political and financial impetus for the Cape to Cairo line had run out. In the interwar period Britain was preoccupied with perilous international relations as well as economic instability. The railways that did exist in British-ruled Africa had largely been built with private money and, with a global downturn, along with the absence of colonial zealots like Rhodes or Kitchener (killed at sea in 1916), there was little hope of pushing such a grandiose project through. With the eventual break-up of the Empire after the Second World War, a cross-continental railway dropped off the agenda completely. The new, independent African republics had much more pressing problems.

In recent years there have been tentative signs that the idea of a Cape to Cairo rail link could be reborn, though not, as Rhodes had hoped, overseen by the British. In the 1970s the Chinese helped build a brand new railway line between Tanzania and Zambia which connected to the railways running down to the Cape. And, as of 2010, plans are afoot to connect up the line in Uganda with the one in the Sudan. It is to be largely funded by a German company, a fact which would surely have Rhodes turning in his grave.

35

FROM RUSSIA TO
AMERICA BY TRAIN

At the turn of the twentieth century, in the same era that Rhodes was trying to connect up railways in Africa, another even more audacious project was being considered – a transport link that would connect five continents and allow people to travel from Europe to America without ever having to take to the perilous high seas.

The key to this idea is the 53-mile-wide expanse of water known as the Bering Strait just south of the Arctic Circle. While remote and suffering from an inhospitable climate, this narrow gap at the very top of the Pacific Ocean is all that separates Russia on one side and the USA on the other. The first to identify the possibility of linking Russia and America via a railway was William Gilpin, a bombastic ex-soldier and the first governor of the US state of Colorado. He was one of the nineteenth century's most pugnacious nation builders. In 1846 he had said: 'The untransacted destiny of the American people is to subdue the continent – to rush over this vast field to the Pacific Ocean … and shed blessings around the world.'

In Gilpin's lifetime railways helped open up North America's wild west to trade and migration, and he felt that trains could help connect the 'new world' to the 'old world' too. In 1890 he wrote a book called *The Cosmopolitan Railway:*

Compacting and Fusing Together All the World's Continents. In it
Gilpin spelled out how a rail route could be forged across
America to Asia and Europe, making long journeys by ship
across the Atlantic unnecessary. Thus one could 'transcend the
disharmony of world geography' and 'so bring together and
intermingle all the people of the earth as ultimately, in a great
measure, to obliterate race distinctions and bring about a uni-
versal brotherhood of man'.

First a railroad would be built across the icy wastes of Alaska,
the US territory which had been purchased from the Russians
in 1867, to the Bering Strait (named after Danish explorer
Vitus Bering who crossed it in 1728). A similar railway on the
other side would bring travellers to the Russian coast at Cape
Dezhnev, the most Eastern point of the Asian continent. Rail
passengers would then be able to transfer across the Strait on
a rail ferry. Eventually this would allow travel by rail between
Asia, Europe, Africa and both North and South America.
Being a good Coloradan, Gilpin saw his own state's capital,
Denver, being a new transport hub as travellers used the new
link to shuttle between the five continents.

Gilpin was killed after a collision with a horse and buggy in
1894 before he could do anything practical to further the idea
of a Bering Strait crossing. But he'd started something, and
at about the same time gold was being discovered in Alaska,
creating new enthusiasm for plans to improve transport links
in the region. In 1892 Joseph Strauss, the man behind San
Francisco's Golden Gate Bridge, drew up designs for a bridge
to span the Strait. But they never got off the drawing board.
His critics said a bridge would be too difficult to build and
dangerous for traffic in such a cold climate.

In 1902, a British explorer called Harry de Windt attempted
to make a 18,000-mile overland journey from Paris to New
York via a partly frozen Bering Strait with huskies, to survey
whether a railway link would work. When the weather played
havoc with his plans, he'd had to catch a lift in a ship to cross the
Strait instead. De Windt called it the worst journey of his life.
But on his celebrated arrival in the Big Apple, he said that his

experience of the conditions told him only a tunnel, for what he called an All-World Railway, would work. He explained: 'As to the possibilities of the passage for railroad travel I am much more firmly convinced than I hoped to be. I fully believe that the road will be completed and in operation within 12 years. The Bering Strait can be tunnelled!' De Windt said that the Diomede Islands, which lay half way across the Strait, could be the site of ventilation shafts for a rail tunnel. He got as far as an audience with President Theodore Roosevelt about the idea, but no further.

In 1905 a French scientist called Baron Loicq de Lobel decided to approach the Russian ruler Tsar Nicholas II with a $300 million plan for a tunnel and connecting railroad. Acting as a promoter for American businessmen Edward Harriman and James Hill, he imagined: 'No more seasickness, no more dangers of wrecked liners, a fast trip in palace cars with every convenience'. The proposal was to build a tunnel, and a 2,500-mile railroad from it, to link up with the new Trans Siberian railway which was being pushed east to west across Russian territory through 5,000 miles of wilderness – an incredible feat for a country still steeped in poverty. For the Tsar, who had just lost a war against the Japanese, the tunnel had the advantage that it might strengthen the country internationally. In 1906 it was reported that he'd given De Lobel's scheme the go ahead.

The idea of crossing the Bering Strait still had many critics in America. The engineering challenge would be significant. As one journalist observed, the tunnel would have to be much longer than the longest tunnel which then existed – the 12-mile Simplon Tunnel in the Alps. The *New York Times* wasn't convinced, saying that, at this point, building a tunnel was: 'about as practical as a plan to colonise the dark side of the moon.' De Lobel needed more than the Tsar's say so to proceed with what he called his Trans-Alaska-Siberian Railway; he needed a formal agreement between the sovereign powers involved.

Despite the apparent agreement of 1906, the next couple of years saw the Tsar and his government blow hot and cold over the strategic merits of the tunnel. The Tsar, it was said, wondered what would happen to the tunnel in a time of war (echoing fears in Britain over the first Channel Tunnel). Not only was the Tsar apparently changing his mind, the money markets in America weren't excited by de Lobel's plan and sufficient funding doesn't appear to have been forthcoming. By the end of the decade, the Bering Strait crossing was lost amid the growing mood of hostility between the great powers.

It was to be some time before anyone realistically considered the idea again. But the outbreak of the Second World War briefly gave it new life. The US government was worried about the possibility of a Japanese invasion through Alaska. So they gave orders for the construction of the Alaskan Highway, a road linking the territory through Canada to the US. It was completed in 1942 and though the highway didn't go as far as the Bering Strait, talk of a tunnel grew again, especially since the Soviet Union was now an ally. This time it was for a subterranean road rather than a railway.

The Cold War was soon to close the door to a link once again. But since the break up of the Soviet Union there has been renewed discussion of joining the two continents with a tunnel. Academics have suggested that using it for both transport and oil would be economically beneficial. Even former Russian president Vladimir Putin has expressed his support for such a project. At no more than twice the length of the Channel Tunnel, the technology certainly exists to get it done. And a Bering Strait crossing is certainly more feasible than a tunnel across the 3,000-mile Atlantic which would achieve the same purpose of connecting the Americas with Europe, Asia and Africa. Such an idea was once suggested by Michel Verne, son of the explorer Jules Verne. In 1895 he wrote a piece envisaging how, one day, steam-driven fans would propel trains through a 3,000-mile Atlantic tunnel at 1,000 mph.

Huge obstacles would lie in the way of such a tunnel. But there's a long way to go before regular traffic starts passing underneath the Bering Strait either. As it stands the infrastructure doesn't exist to get people to the watery impasse even if there was a tunnel. And long-distance flights have now gone some way to achieving Gilpin's original vision.

36

BRITISH RAIL'S FLYING SAUCER & THE 'GREAT SPACE ELEVATOR'

It would certainly have been difficult to persuade anyone who had the misfortune to consume a British Rail sandwich in the 1970s that the network's service could ever be out of this world. So it comes as something of a surprise that just a year after the first moon landing in 1969, a patent with the number 1310990 was lodged by British Rail for a space vehicle. The man behind this bemusing, flying-saucer-shaped craft was scientist Charles Osmond Frederick, who worked at the British Rail research department in Derby. He provided detailed drawings for a 120ft vehicle in which passengers would be ferried into space in a compartment above engines powered by 'a controlled thermonuclear fusion reaction ignited by laser beams'.

It wasn't an April Fool's joke. When the patent was rediscovered in later years it emerged that British Rail had put its name to the idea simply because it had been done by an employee. They didn't, so they maintained, have any immediate plans for a branch line into space. Scientists who have looked at the blueprint since say the technological worthiness of such a craft is questionable, to say the least. The episode does, however, prove the enduring romance humans have with attempts to get beyond the Earth's atmosphere.

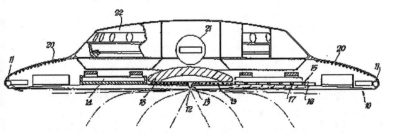

A patent for a flying-saucer-style craft, lodged by British Rail in 1970.

The first record of anyone actually trying to do so has been credited to a Chinese man called Wan Hu. History logs him as the world's first astronaut. Well, attempted astronaut. The story goes that in around 1500 he attached forty-seven rockets to a chair and had them lit, simultaneously, by his servants. Sadly the resulting explosion left no trace of Wan, though he does now have a crater named after him on the dark side of the moon. He was, as it turned out, on the right lines. We now know that it was the pioneers of rocket technology that finally enabled man to get into space. The first sub-orbital trip was made by a German V-2 rocket during World War Two, and the first human being into space was Yuri Gagarin in 1961 aboard Vostok 1.

Widely acknowledged as a leading light of early rocket science was the Russian Konstantin Tsiolkovsky, born in 1857. A reclusive figure, he spent most of his life as a teacher occupying a remote log cabin while working on his scintillating scientific ideas. Tsiolkovsky never built an actual rocket. But his theories and formulas did lead the way for others to do so in the early decades of the twentieth century. His 1903 work *The Exploration of Cosmic Space by Means of Reaction Devices* explained how a liquid-fuelled rocket could achieve orbit in space and, late in life, he was given a government pension to work on his theories, which included the idea of using multi-stage rockets to get into space.

By the end of the 1920s his importance in the field was being recognised. Hermann Oberth, a famous German scientist, wrote to Tsiolkovsky saying: 'You have ignited the flame, and we shall not permit it to be extinguished; we shall make every effort so that the greatest dream of mankind might be fulfilled.' Tsiolkovsky, who died well before the rockets he foresaw actually reached space, even looked ahead to a time when new technology would have to replace them. He predicted that the business of launching rockets would be extremely expensive and thus came up with an idea to tackle the problem: the 'great space elevator'.

The idea still seems far-fetched today, bringing to mind visions of Willy Wonka setting off for the inky depths with Charlie Bucket in Roald Dahl's 1972 book *Charlie and the Great Glass Elevator.* But it was the newly built Eiffel Tower and its brand new lifts speeding visitors to the top that convinced Tsiolkovsky that a version to take man into space could be more than a fictional fantasy. The Parisian landmark led him to formulate the idea of a tower, or cable, which would be tethered to the planet, extending 22,300 miles upward. In his 1895 work, *Dreams of Earth and Sky*, he spelled out how, at the top, there would be a 'celestial castle' – a kind of space station. This would be held in a geosynchronous orbit – one where the speed of the orbit matches the rotational speeds of the earth. This way it would be far enough up and stable enough so that spacecraft could dock with it. Human space-flight would become easier, more affordable and so promote the exploration of the cosmos.

Of course he knew his tower could never be constructed in his lifetime. It would need to be made out of something stronger than the iron which makes up the Eiffel Tower. Even steel wouldn't do. But Tsiolkovsky's space elevator idea may not have been simply pie in the sky. As Tsiolkovsky pointed out, the current expense of ferrying things and people into space using rockets isn't sustainable and so space scientists are looking at the space elevator idea all over again. They think the technology to build one is not only possible but could become reality in just a few decades. The basic science works and new super-strength materials like carbon nanotubes are being investigated while a leading NASA scientist, David Smitherman, has even said that in fifty years' time the technology should have developed sufficiently to allow construction. He says the idea of a space elevator is no longer science fiction. All that is needed, it seems, for Tsiolkovsky's idea to become a reality is the will to do it.

Tsiolkovsky himself thought that human exploration of space was essential for human life to survive. He said: 'The finer part of mankind will, in all likelihood, never perish – they will migrate from sun to sun as they go out. And so there is no end to life, to intellect and the perfection of humanity. Its progress is everlasting.'

37

WHY THE BATHYSCAPHE WAS THE DEPTH OF AMBITION

On six occasions, human beings have visited the moon. And the number of people who have spent time in space is measured in hundreds. Yet while man's quest to conquer other worlds continues, we still remain largely ignorant of what lies beneath on our own planet. Just two people have been to the deepest reaches of the earth's most inhospitable ocean canyon, almost 7 miles underwater. Doing so required a hugely expensive, highly effective contraption that reached parts that other vessels could not.

Bathyscaphe (or bathysphere as it is sometimes called) combines two Greek words in a construction that means 'deep boat', but this does the invention little justice. For a bathyscaphe called the *Trieste*, a magnificent deep-water vessel designed by Swiss father and son team Auguste and Jacque Piccard, was to reach the bed of the Mariana Trench in the western Pacific. Any craft risking such a journey, the deepest part of Earth's seabed, must withstand pressure of 17,000lbs. Any structural weakness means certain death to those aboard.

Having already been to the edge of space in a self-designed balloon, Auguste Piccard's attempt to go deeper than anyone has done before, or since, was to be his second major adventure. In 1932 he ascended 16,650m to study cosmic rays from

The *Trieste*, a diving research vessel which visited the deepest part of the Earth's oceans.

the stratosphere. Turning to the seas with his son, also an accomplished physicist, and having secured funding from the Italian city of Trieste – hence the name of the vessel – they built their bathyscaphe. And, in progressively more challenging dives, it worked.

More diving bell than boat, with an appearance combining submarine with torpedo, the bathyscaphe's first significant journey took it 3,099m to the sea floor off Ponza, Italy, on 1 August 1953. Just one scientific achievement resulted: the Piccard's proved they could get to the seabed and back and survive. Unfortunately, despite its success, this first bathyscaphe was later destroyed in a squall. A second vessel fared better, while the third, the *Trieste*, was to go down in US naval history. Fifteen metres long with protective walls 13cm thick, and propelled by electric motors, it resurfaced by releasing ballast pellets from magnetic hoppers, and could survive the deepest reaches.

The greatest challenge to deep sea exploration isn't so much the technology as the cost – and the politics. Needing endless amounts of money to keep the project afloat, the Piccards turned to the United States where the Navy enabled them to improve their vessels. By 1960, they were ready to face their toughest test: the world's deepest ocean canyon. At 10,918m, Mariana Trench is 20 per cent deeper than Everest is high, but while Everest has the advantage that on clear days the views are magnificent, it's a different story on the seabed. After twenty minutes peering out of the tiny windows into the darkness, the Piccards had seen all there was to see: one seabed-dwelling flatfish that they were unable to photograph. Proving only that the bathyscaphe was up to the job, they returned, happy but empty handed, to the surface. And that was that. But for this one historic attempt, man has never returned to such depths.

Back on dry land, US military focus was changing. The Navy thought that, by and large, going to the bottom of the sea to see nothing much was a poor use of precious resources that could be better spent on nuclear missiles. Vice Admiral Hyman G. Rickover, known as the 'father of the nuclear navy' for directing the service towards nuclear subs after the Second World War, put a stop to further *Trieste* dives.

That appeared to be the end for the *Trieste*. But in a judicious, if tragic and ironic twist, it wasn't long before Rickover was forced to call the vessel back into action. On 10 April 1963 one of the nuclear-powered submarines for which he was responsible, the USS *Thresher*, sank. Somewhere between 400–800m below the surface, the *Thresher*'s hull had collapsed under the tortuous surrounding pressure. 129 crew lost their lives as seawater flooded in and the *Thresher* exploded. If this looked like a serious situation in itself, it was worse than it seemed. A nuclear reactor was out of control, out of reach and at the mercy of the elements. A shocked Rickover tried to assure a sceptical world that a radioactive leak was impossible.

By this time the *Trieste*, still a US Navy ship, was lying redundant in San Diego. Its task now: locate the *Thresher* and recover sufficient parts to discover what really crippled it. The *Trieste* was fitted out with a new, untested, mechanical arm for a recovery mission and delivered on board the USS *Point Defiance*. In June, in an area thought to contain the wreckage of the *Thresher*, the *Trieste* began the first of a series of dives. Three men crammed into the vessel for each trip to the ocean floor. It wasn't easy, and a bathyscaphe isn't a comfortable place to be at the best of times. When you're encased in a hard metal shell with the possibility of landing on a nuclear reactor or, no less frightening, a nuclear missile, nerves are tested all the more.

Locating the wreckage took several attempts. On the first, small sound generators called 'pingers' that were meant to guide the *Trieste* into place, failed, directing the crew to the wrong area. On the second attempt, the pilot steered the bathyscaphe into the mud on the seabed, where it became firmly stuck. After jiggling about for half an hour, the vessel scraped itself loose and hurtled away. On the third attempt, the crew found a plastic shoe cover used by mariners in the reactor compartment of a nuclear sub. At last, they were in the right place, although a fourth dive located nothing. On the fifth attempt, the gyrocompass broke, the starboard propeller malfunctioned and the navigation system closed down, leading to a tow back to Boston for repairs ahead of any further attempt. This wasn't looking good for a cutting-edge machine that had been given a second opportunity to show what it was made of.

Then, on 29 August, success came at last. Although its 4,500 watt light could peer only 50ft into the blackness, the bathyscaphe's underwater camera located twisted metal, including a piece of pipe marked with the *Thresher*'s name. Using the *Trieste*'s new mechanical hand, the debris was hauled in.

Since those heady days of the 1960s, bathyscaphes have seen little action. Subsequent ones made by the Chinese reach only a third of the depths achieved by the *Trieste*. The US Navy has a submersible device called *Alvin*, made in part out of syntactic, buoyancy foam made by General Foods in the same factory where they produce breakfast cereals. But it can only reach depths of 4,500m, a snip of that achieved by the *Trieste*. Despite notable ups and downs, however, it has been infinitely more successful. It located a lost hydrogen bomb off the coast of Spain in 1966 and was attacked by swordfish a year later, forcing an emergency ascent to the surface where the fish was removed from the skin of the vessel and cooked for dinner. The *Alvin* sank the following year but was eventually recovered from the seabed and refitted, going on to discover vast colonies of 3m-long tube worms, a species previously unknown. Although coming up to fifty years old, it remains the planet's most sophisticated aqua-research vessel.

Nuclear submarines still steal through the watery darkness and occasionally get into terminal difficulty, despite the tragedy of the *Thresher*. Today, however, tourism and warfare pay more attention than science to what goes on in the very reaches of the earth. The world has its first tourist submarine, the *Auguste Piccard*, built by Jacques Piccard in honour of his father, and a larger one, the PX-44, for sizable holiday parties. But ultimately, anyone who wants to intimately understand the topography of the ocean floor will be disappointed. Man has better maps of Mars and the moon than the seabed. And there are no plans to change that.

38

THE PERPETUAL MOTION MACHINE

One day as Greek mathematician Archimedes sat in his study in the third century BC, a courtier of King Hiero II appeared with a newly cast crown, purportedly made of pure gold. But the king was not so sure. Could Archimedes tell whether a corrupt goldsmith had bastardised his crown with silver?

As with many of the best ideas through the ages, the answer came to Archimedes in the bath. The water level, he noticed, rose when he lowered himself into the tub. Ever the mathematician, Archimedes concluded that: 'a body immersed in fluid exerts a buoyant force equal to the weight of the fluid it displaces.' By placing the crown in the bath he was thus able to work out the density of the gold in the crown, by dividing its mass by the volume of water it displaced. History doesn't tell us why he felt compelled to get into the bath with the crown, nor whether he was wearing it at the time. But the answer came to him in a flash. Excitedly running through the streets naked shouting 'Eureka', he announced the result: the crown was indeed tainted by silver. To this principle of water displacement he gave the humble name Archimedes.

For his next trick, Archimedes was to create a screw to help Greece build the largest warship in naval history to date, *The Syracusia*. Because classical Greek shipbuilding wasn't quite

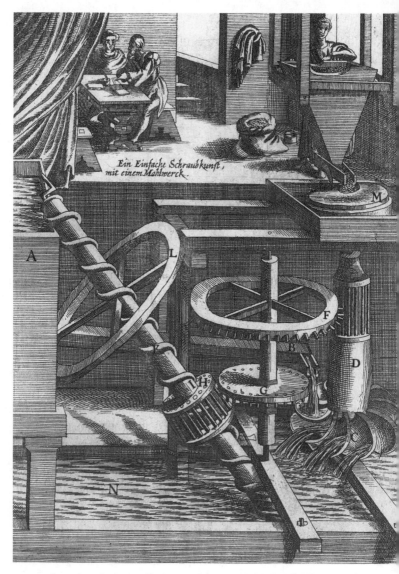

Ein Einfache Schraubkunft,
mit einem Mahlwerck.

A seventeenth-century illustration of Robert Fludd's perpetual motion machine.

up to scratch at the time, the battleship took on water on each voyage, which is never a reassuring prospect for a ship. Archimedes' Screw solved the problem. Through a revolving screw in a cylinder, water could be raised endlessly: the water turns the screw that turns the pump, almost but not quite creating its own power to bail out any incoming seawater. Even better, the screw allows water to run uphill. All in all, quite an achievement for a rotating screw.

As the centuries passed, scientists wondered whether the principle could be improved so that machines could work without an external power source. If they could, and if they could do so endlessly, man could put his feet up and let machines run the world. Since then the quest for perpetual motion has driven inventors variously mad, broke or to suicide. And until someone finds a way to break two fundamental laws of physics, perpetual motion will remain impossible. The good news is that that's exactly the kind of challenge that scientists adore.

Renowned inventors, including Richard Arkwright, who came up with the spinning jenny, and father of the railways George Stephenson, worked on the margins of the perpetual motion concept. Others went full blast to design a machine that, without any external energy source, would produce enough power to keep on going forever. Many were men of religious conviction, and a small number were threatened with ex-communication for such heretical experiments. But they believed that if nature could exhibit perpetual movements – every morning the sun rose, every evening it set; twice a day the tides ebbed and flowed – then man could surely emulate it.

While the first kind of water motor was the Buddhist praying wheel, in which prayers are fastened to a wheel and rotated using water power, serious attempts to keep a machine in action in perpetuity didn't emerge until the Renaissance – and not before Bruges mathematician Simon Stevinus tried to put a dampener on the whole enterprise by tying fourteen balls to an endless piece of cord, suspending them from a triangular frame, and through the law of equilibrium proving that perpetual motion could not exist. Shortly after, at the beginning of the seventeenth century, Florentine mathematician Galileo Galilei fastened one end of a cord to a nail in a wall and swung a heavy ball from it. Although he almost inadvertently invented swingball, he unfortunately couldn't keep the exercise going for long. Perpetual motion was proving evasive.

It fell to an English man to develop the concept further. In 1618 aristocratic courtier Robert Fludd developed a water-mill than didn't need a flowing river to run it. The idea was to provide the miller with all the power he needed to grind grain, all day, every day, forever. But Fludd was a philosopher and religious fundamentalist, as well as a doctor, alchemist and inventor, so first he had to define what 'forever' meant. This was not simple. By definition, perpetual motion machines must move in perpetuity. But does that mean for a few years until it breaks down, until God calls time on man, which was the prevailing Christian view, or until the universe caves in on itself, which hadn't even been thought of? Time is a tricky physical and philosophical concept, and designing a machine that would move till its end made the brains of even the smartest inventors throb.

All of this concerned Fludd, a very serious Jacobean thinker. Often insightful, frequently a fantasist, he was, in many respects, dangerously ahead of his time: claiming, like Galileo, that the sun, not the earth, was the centre of the universe – a theory that had the Italian, but not Fludd, hauled up in front of the Pope. He was also convinced that blood moved in a circu-lar motion around the body, something not broadly accepted in seventeenth-century England. Divinity informed much of Fludd's philosophy too. Lightning was the will of God, he insisted, and those who got struck by it got everything they deserved. His theory of disease was somewhat unusual for a man of medicine: a wind controlled by a powerful evil spirit excites lesser evil spirits in the air which then infiltrate the body – and if you don't get to the apothecary fast – ideally the one he ran from his home in London's Fenchurch Street – woe will almost certainly betide you.

With so much going on in Fludd's mind, what with the apothecary to run, the king to pacify occasionally and learned books to write, he sometimes found little time to devote to his lifetime's abiding interest: wheat. But whenever he could, he experimented with the cereal, writing extensively on what he called 'The Excellency' of the crop. His studious interest in milling turned into a series of exquisite drawings of a closed-cycle mill using a form of Archimedes' Screw. Water turned a wheel to power a pump that would cause the water to flow back over the wheel. This would then power the pump to cause the water to flow back over the wheel, and so on and so on until everyone got dizzy. And although all this detail was only on paper, the physical laws that would invalidate the concepts were still a long way from being understood.

Although he wasn't the first to design a perpetual motion machine, Fludd was the inventor of the closed-cycle mill. In the following decade, Italian philosopher and alchemist Mark Antony Zimara designed a perpetual motion machine in the form of a self-blowing windmill, in which two or more bellows puff air produced out of nowhere at the arms of the windmill. He was a little vague about how to produce the energy, leaving it 'to the ingenuity of the maker', but was convinced that the arms would rotate forever. By 1653 Edward Somerset, the Marquis of Worcester, who some believe created the first practical steam engine, had a plan 'to raise water constantly, with two buckets only, day and night, without any force but its own motion'. After several long days and nights he eventually concluded it wouldn't work. Over in Germany, Ulrich von Cranch's 1664 machine dropped a ball on top of a paddle-wheel which turns the wheel as the ball falls, rotating an Archimedean Screw. This then carries the ball back to the top of the device when it reaches the bottom. Then the ball falls and the process starts all over again. For the purposes of efficiency, von Cranach used more than one ball and spent hours calibrating the device. But it never overcame the laws of physics to keep going without someone dropping a ball now and again. And in 1686, Georg Andreas Böckler produced numerous ideas for perpetual-motion machines, including 'self-acting mills', in which cups or buckets on an endless rope re-deliver water to a wheel, rotating the wheel and moving the cups in a virtuous loop. It worked perfectly on paper, but nowhere else.

The natural laws of thermodynamics, still almost 200 years away from being defined, prevented the seventeenth-century perpetual-motion machines – and all subsequent and, in all probability, future ones – from working. The first and second laws of thermodynamics say, respectively, that energy can be neither created nor destroyed but only changed in form, and that it is impossible to make a machine that doesn't waste some energy, however small. In practice, the second law of thermodynamics means that any friction created by the wheel of a perpetual motion machine turns into heat and noise, thus losing energy. And if the machine for whatever reason doesn't cause friction, it will still need more energy than it produces to keep the wheel itself going, thus breaking the first law. In short, perpetual motion machines are impossible, yesterday, today, and although one should never say never, in all likelihood forever. The nearest we get today are nodding dogs in the backs of cars, and only then when the vehicle is in motion.

39

DISASTER

BRUNEL'S NOT SO
GREAT EASTERN

With two record-breaking steamships behind him, engineer, entrepreneur and showman Isambard Kingdom Brunel needed an idea bigger, better and bolder to continue pushing the boundaries of engineering. But the result, his next ship the *Leviathan*, later known as the SS *Great Eastern*, proved ruinous. Arguably, it killed him.

A monster of a ship conceived in 1852, Brunel's *Leviathan* would demolish records by the score. At 211m long and weighing in at 22,500 tons, its volume would be six times larger than any existing vessel. It would displace more water and sport more masts, sails, furnaces and crew. In a fair wind, it would steam its way non-stop across 22,000 miles of ocean; almost enough to get the whole way round the globe without refuelling. And by travelling great distances faster and providing comfortable, air-conditioned cabins deep within the hull, it would also be a commercial success. 'I never embarked on one thing to which I have so entirely devoted myself, and to which I have devoted so much time, thought and labour, on the success of which I have staked my reputation', Brunel wrote later. This was prescient; for by the time the ship sailed, his reputation was in tatters.

From start to finish, the *Leviathan* appeared cursed. Firstly, the Eastern Steam Navigation Company, which was talked into ordering the ship from the ever-persuasive Brunel, was left with nothing to use it for when it lost a government contract (Brunel's prospectus was nothing more than 'gross bait', rued the chief constructor of Woolwich dockyard). Then shipbuilder John Scott Russell, delighted when the famous engineer accepted his remarkably, possibly cynically, low bid for construction, began to cut corners to meet the ludicrously tight budget. But most significantly, the choice of shipyard proved disastrous for a vessel of such magnitude.

Early in the project, Brunel had been forewarned of launch difficulties by the Institute of Civil Engineers. The ship could never launch safely in the Thames at the Isle of Dogs, they cautioned – the width of the river at this point being not much greater than the length of the ship. But securing dock facilities with a launch site sufficient to handle the world's heaviest ship would add to costs. Ignoring the civil engineers' advice, Brunel built the *Leviathan* at a dock from where it would be unable to launch safely. Nonetheless, as the shell began to take shape, its magnificence could not be doubted. The double hull – with 30,000 iron plates, each uniquely shaped, stamped with a number and then fitted together in the fashion of a giant jigsaw puzzle – was the kind of engineering for which Brunel was famed. The six giant masts were each given the name of a day of a week, from Monday to Saturday, allowing the crew to quip that they would never have a Sunday at sea.

But all was not well. Scott Russell's undoubted skills as a shipbuilder proved insufficient to compensate for his woeful underestimation of the complexities of the *Leviathan*'s construction. As costs spiralled, his debts began to mount, until eventually suppliers were unwilling to work with him. Brunel didn't help, 'throwing his weight around' and becoming 'critical, petulant and downright rude in his dealings with him' according to historian George Emmerson. Because Brunel only paid Russell on the weight of iron erected, and a quarter of that payment was in worthless shares, bankruptcy loomed. At one point, with Russell's creditors within their rights to seize the ship and Brunel believing he had extracted all the value he could out of him, Brunel chivvied the Eastern Steam Navigation Company into accepting early delivery of the ship. A quarter of the hull remained incomplete, fitting out was some way away, and there was still the problem of that looming launch in the Thames. If that wasn't bad enough, disaster was to follow.

Tragedy had already visited the *Leviathan*. A work boy, falling head first onto a spike, died horribly. And working away in a watertight and, as it turns out, unfortunately airtight cell between the gaps of the double hull, one riveter slowly suffocated, his yells going unheard as a thousand colleagues banged away in adjacent cells. His death, it is said, brought the bad luck to the ship, as one visitor discovered when his own head was crushed in as he looked around.

But industrial accidents, even fatal ones, did not derail construction unduly in Victorian England, and before long Brunel decided the time had come to launch. To overcome the problems of the site, this would be done sideways and at an angle that would lift the ship's bow 12m. Taking advantage of a high tide on 7 November 1857, and with the ritual crash of champagne against its side, the *Leviathan* launched. Anticipating problems, Brunel was eager to keep crowds away. But you can't launch the biggest ship the world has ever seen, on which thousands of workers have been employed, and which has generated such publicity (not all of it good) and hope that nobody notices. The enterprising Eastern Steam Navigation Company, anxious to recoup some of its investment quickly, didn't help by selling 3,000 tickets.

The launch was calamitous. As restraining chocks were knocked away and hydraulic presses began the process of getting the ship into the water, its impressive but inadequate chains snapped, leaving the *Leviathan* lurching precariously on the launch pad. Although Brunel had loaded gallons of water to the front of the ship to redress an imbalance on the two cradles of the pad, this only added several hundred tons more weight and caused havoc with the manoeuvre. As it creaked and groaned, it looked like the ship would topple. Workers fled in terror. Spectators screamed as the hulk slid towards the water. But then, after just a few metres it shuddered to a halt. The damage was done. In the space of just a few minutes, two people had died and several were seriously injured. The sideways launch – and Brunel himself – was condemned by *Mechanics Magazine*. 'An unnecessary display of self-confidence', it admonished.

Two months later, Brunel tried again, but accepted that the Isle of Dogs location had been a bad idea all round. Even for the famous engineer, choosing another way to launch the *Leviathan* was a puzzle too far. Fellow inventor, the even more influential Robert Stephenson, by this time preoccupied by the process of dying, rose from his bed, got into his dressing gown and slippers and reached for his protractor. More powerful hydraulics would be needed if this ship was ever to set sail, he said. Yielding 3,500 tons of force in addition to the 100 tons of gravity, the extra hydraulic power worked. On the second attempt on 31 January 1858, the *Leviathan* shamefully slipped into the Thames, tearing up the bed of the river as it did so, with no sign of the large crowds that had risked their lives before. £1 million had been spent by this time, including £170,000 on the launch, and the ship had still to be kitted out. The Eastern Steam Navigation Company was almost broke, but at least now it could begin to make money.

On 9 September 1859 on its way to Weymouth, from where it would begin its maiden voyage to New York with paying passengers, disaster struck again. When the ship reached open waters an explosion tore through it, bringing a funnel smashing onto the deck, shattering cut glass mirrors and tossing furnishings around. In the boiler room, five crew were killed. The ship limped to Weymouth, where it anchored for repairs and for the bodies to be removed. The news rocked Brunel. Already ill, he took it badly. Not long after, he was dead, beating his friend Robert Stephenson to the grave by four weeks. He was 53.

It was nearly a year before the *Leviathan*, now called the SS *Great Eastern*, was shipshape again, but its reputation went before it. Just thirty-five passengers paid for the first voyage on 14 June 1860, and were looked after by 418 crew. To the dismay of the passengers who by that time had boarded, the captain put back the sailing by three days because the crew was drunk. The ship's curse continued to strike. The captain who had taken her into Weymouth after the explosion, William Harrison, later capsized in one of the ship's boats in a squall near Southampton and drowned. On its second voyage the following year, the boat taking passengers to the ship off Milford Haven ran aground and they had to be rescued by other boats in the vicinity. On the third trip, a gale rolled the ship, smashing the starboard paddle wheel beyond redemption, causing the port paddle to be lost completely and damaging the rudder. Passengers were injured. For three days the ship drifted at sea, out of control. It reached the south coast of Ireland, where the harbour master at Queenstown refused the ship entry because the captain was unable to control it. Once again, passengers had to take to small boats to get ashore.

It was the end for the SS *Great Eastern* as a passenger ship, but there was some modest success ahead. After being converted into a cable-layer, it installed the first transatlantic telegraph cable in 1866, before becoming a floating music hall. Brunel's *Great Babe*, as he had called her, was finally broken up for scrap in 1889 – although being so big, this took eighteen months. *Mechanics Magazine* summed up the venture: 'No failure of his ever did so much to lower the reputation of English engineers as the launch of the *Leviathan*.'

40

BESSEMER'S
ANTI-SEASICKNESS SHIP

'Few persons have suffered more severely than I have from sea-sickness.' So wrote Sir Henry Bessemer, the renowned inventor and industrialist, in his autobiography. On a journey across the English Channel from Calais to Dover in 1868 he'd been so ill that, on his arrival back at his London home, his doctor had attended to him throughout the night. According to Bessemer the doctor had even administered a small amount of prussic acid (actually a poison), to try and cure the malady.

Some seven years after that awful journey Bessemer was at sea once again, this time travelling from Dover to Calais on board a ship named after him – the SS *Bessemer*. The ship had been built with a very specific purpose. It was designed to solve, once and for all, the common ailment of seasickness which had long afflicted millions. Even that hero of the Battle of Trafalgar, Lord Nelson, had been a sufferer.

Sir Henry Bessemer is best known for coming up with a way of making cheap steel. His Bessemer Converter was one of the great machines which helped power the industrial revolution and, later, made skyscrapers possible. By the time his huge paddle steamer the SS *Bessemer* put to sea, the Hertfordshire-born genius had become internationally famous and very rich. In his lifetime Bessemer had come up with scores of patents covering everything from pencils to artillery shells. Many of his ideas were unassailable triumphs. His ship that would leave its passengers free from seasickness would prove to be a rather more brittle proposition.

It was Bessemer's success that took him to sea, as he travelled abroad on business. But he loathed it. Following that dreadful bout of seasickness in 1868 he decided to do something about the problem which he knew affected so many. His brainwave was to fashion a 'swinging saloon' inside a ship that would be set on gimbals, or pivoted supports. Weighted underneath it would move independently of the ship's hull, kept horizontal with the help of hydraulics. This way, he believed, the saloon would be free from the rolling motion of the ship, which was the chief cause of the sickness which afflicted him and others.

Bessemer started work on a large model in his own garden at Denmark Hill in South London. The location was chosen partly because Bessemer couldn't bring himself to test his invention anywhere near the sea. Eventually his model saloon was big enough for people to sit inside. This way he could demonstrate the idea to his peers. Using a steam engine, Bessemer replicated the movements of a ship at sea and, by all accounts, was able to make his saloon stay level. The inventor reported that his fellow engineers had proclaimed his experiments a complete success.

A firm, the Bessemer Saloon Steamboat Company, was soon set up to oversee the building of a cross-Channel steamer that would be specially adapted to accommodate his contraption. At 350ft long, the ship would be much bigger than the existing steamers. This length, it was felt, would reduce pitching at the front and back of the ship. The saloon would help reduce the rest of the motion aboard to virtually nothing.

In a letter to *The Times* in 1872 the ship's respected architect, E.J. Reed, who had once been chief constructor of the Royal Navy, explained how he saw the SS *Bessemer* working:

> I do not put her forward as a perfect remedy for sea-sickness in all cases, although I think she will be found a sufficient remedy in the Straits of Dover. Her advantages seem to me

to be that she will be large enough herself to escape all but very small movements as regards lifting bodily and pitching. The moderate pitching which she would otherwise experience will be diminished by the low ends, and what remains of it will scarcely be felt at all in the centre saloon. The rolling of the ship, which is the only remaining movement of importance, will be perfectly neutralised by Mr. Bessemer's hydraulic arrangements.

In the same year the scientific journal *Nature* agreed that the project was surely on course for success with Bessemer and Reed on board: 'The association of those names is in itself a sufficient guarantee that the idea will be carried into execution with complete security as respects the safety of the passengers and the seaworthiness of the ship, and a full knowledge of the scientific principles involved.'

Convinced that his invention was a masterstroke, Bessemer ploughed much of his own money into the company and the construction of the ship. No expense was spared on the fixtures and fittings of his Saloon either, which measured 70ft by 30ft wide and 20ft high. It featured extravagant carved oak columns, gilt panels and hand-painted murals. By 1875 the SS *Bessemer* was delivered from the shipyard and its first public voyage was set for 8 May.

Just to be on the safe side, in April the ship made a trial run across the Channel. It didn't go to plan. The ship had bumped into the pier at Calais trying to dock, wrecking one of its own paddle wheels. With the much-publicised public launch approaching the ship was hastily fixed. These repairs meant Bessemer had no chance to finish adjusting the complex mechanism that controlled his saloon. However, instead of delaying the launch, Bessemer decided that, on this occasion, the saloon would have to be fixed rigidly in place for the trip. He'd have no chance to demonstrate just what it could do but at least he could show off the grandeur of the saloon's interior and get people talking. The mechanism could simply be put fully into action on a later run.

After an uneventful passage across the Channel, with its passengers dutifully admiring the saloon's handiwork, the ship was approaching Calais harbour in clement weather. Suddenly disaster struck. The captain had calmly ordered the ship to be moved in one direction – only for it to lurch in another. Bessemer describes how, to his horror, the ship began heading towards Calais pier once again, only this time the collision was to be much worse than the one in April. With a loud, splitting sound the ship ploughed into the structure knocking down its timbers 'like ninepins'. One passenger later admitted, with some shame, that the disaster was met with a great roar of laughter on board. The great man himself wrote:

> I knew what it all meant to me. Those five minutes had made me a poorer man by £34,000; it had deprived me of one of the greatest triumphs of a long professional life, and had wrought the loss of the dearly-cherished hope that buoyed me up and helped to carry me through my personal labours. I had fondly hoped to remove for ever from thousands yet unborn the bitter pangs of the Channel passage.

After the event, Bessemer maintained that he'd not had the chance to finesse his saloon system and put it through a proper trial. But it seems that the ship's Captain Pittock, a veteran of the Channel run, had found the craft built to house it difficult to steer at the slow speed needed for entering a harbour. The weight of the Bessemer Saloon in the middle and its heavy engines at either end couldn't have helped. The failure of the voyage made investors panic and in 1876 The Bessemer Saloon Steamboat Company was wound up.

The SS *Bessemer* itself was scrapped, though its architect, E.J. Reed, eventually had the saloon section moved to his own home in Stanley, Kent where it became a private billiard room. It survived until the Second World War when a direct hit from a German bomb blew up all that remained of Bessemer's floating folly.

Bessemer's reputation managed to recover from the embarrassment of the episode and history has been kind to him.

Today thirteen towns in the USA still bear his name. He might be disappointed that no one since him has successfully come up with a design of ship that entirely can cure seasickness. But he'd probably be thankful that, these days, he would be able to take a flight across the Atlantic to visit all those American towns which share his name, rather than endure agonising weeks at sea.

41

ESCAPE COFFINS FOR THE MISTAKENLY INTERRED

Death is always a nuisance, but failing to die before burial can prove terminal. In the eighteenth and nineteenth centuries the thought of waking trapped in a coffin, deep in a grave, was a very real fear, and, evidence suggests, sometimes a reality. Many are the tales of those prematurely buried, sometimes deliberately as a form of execution, but sometimes accidentally too.

So what could be handier than a coffin that those prematurely interred could climb out of? Taphophobics, those who fear being buried alive, and cataleptics, those suffering paralysis so severe it gives the impression of death, could rest easy with a range of contraptions to bring salvation.

For at that time, death suddenly wasn't always what it seemed. What had previously been a cut and dried matter of the heartbeat stopping was no longer necessarily so. Medical advances, slow though they were, meant that people who up until that point were formally deceased could now sometimes be resuscitated. Heartbeats could return. And the debate raging in medicine, in philosophy and in literature, frightened people to death. In the 1730s Jacques Benigne Winslow captured the spirit eloquently in the title of his book *The Uncertainty of the Signs of Death and the Danger of Precipitate Interments and Dissections*, which related how common it was for death

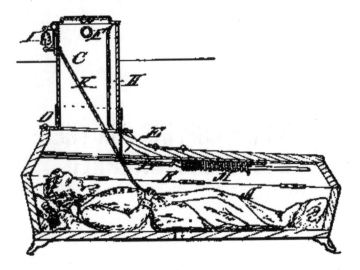

Franz Vester's 1868 patent illustration for an 'improved burial case'.

to be declared prematurely. Even decay of the supposed corpse couldn't be taken as a sure sign of death, according to researcher Charles Kite, since putridity was also a symptom of advanced scurvy. Some accounts suggested that one in ten burials may have been conducted a little hastily. And so the inventions came.

Franz Vester's 'improved burial case' – what most people, dead or alive, would know as a coffin or casket – became the first US patent of its type in 1868. Vester's 'improvement' was primarily a wide vertical tube running up 6ft of earth to the ground above the grave, attached at a 90° angle to the box in which one usually spends eternity. On waking, relieved to be alive and to have had the foresight to prepare for such an eventuality, the corpse would then use a ladder provided in the lining of the tube to ascend to ground level. Anticipating some difficulties in manoeuvring oneself around the coffin, presumably a little stiff from lying around to be able to use the ladder easily, Vester supplied back-up in the form of a bell. By tugging on a cord placed in the hands before interment, the corpse would ring to attract attention in the graveyard above, no doubt to the alarm of mourners at neighbouring graves.

Vester admitted that, in most eventualities, neither the ladder nor the bell would see active service and that, after a suitable period of mourning, people should accept the inevitable. With an early eye on recycling possibilities, he wrote: 'If, on inspection, life is extinct, the tube is withdrawn, the sliding door closed, and the tube used for a similar purpose.' Of its type, Vester's improved burial case possessed a remarkable level of ingenuity not considered by lesser inventors. Several designs included an alarm system of some kind, a bell on the surface being the most usual, but they neglected any way of getting oxygen to the body. Even those who awoke to find themselves peering dismally at a coffin lid, but relieved that at least they could ring for service, would be long dead before help came along.

Before Vester's improved burial case, in 1843 Christian Eisenbrandt registered a coffin-dodging device with a spring-loaded lid. But unlike Vester's pioneering idea, it wouldn't function after burial, which was rather a disappointment if you woke up to the sound of earth being shovelled on top. 'The slightest motion of either the head or hand acting upon a system of springs and levers cause the instantaneous opening of the coffin lid,' says Eisenbrandt's patent application. More Jack in the Box than coffin, there is no record of the device, or Vester's, ever going into manufacture.

So to the grave people continued to go, where presumably they sometimes woke up blinking into the darkness and sensing they might have a problem on their hands. In 1858, the magazine *Notes and Queries* reported how the relatives of a deceased wealthy widow, bearing her to interment next to her beloved who had passed away fifteen years earlier, had a shock at the mausoleum. When the tomb was re-opened 'the coffin of her husband was found open and empty, and the skeleton discovered in a corner of the vault in a sitting posture'. Around the same time, Edgar Allen Poe, eager to repeat the success of his gothic story *The Fall of the House of Usher* in which a man buries his cataleptic sister, rushed out more yarns along the same theme, including the self-explanatory *The Premature Burial*. Readers lapped it up, and even the famous, sane and otherwise rational began to take precautions. US President Washington decreed that his body should be kept above ground for three days before burial, just in case. And impressionist artist Augustine Renoir, who wasn't to die until 1919, was so afraid of premature burial that he had his son instruct doctors to 'do whatever was necessary' to ensure that he was truly dead. But records are sparse to the point of non-existence on the successful deployment of an escapology coffin by anyone formerly declared deceased.

42

CHADWICK'S MIASMA-TERMINATING TOWERS

Cities kill; especially nineteenth century cities, and most especially smelly cities. When people died in their tens of thousands in urban centres from fearful diseases such as cholera, diphtheria and typhoid, the logic was obvious. Smells caused disease which caused death. So if just one thing could be done to stem the rise of lethal diseases in Victorian Britain, nothing would beat eradicating odours, and Edwin Chadwick, leading health campaigner and early adopter of the comb-over, was the person to do it.

Chadwick's credentials for establishing the systems that would improve air quality were outstanding. Unfortunately, in respect of the epidemics striking urban Britain, his assumptions, together with those of many eminent experts in public health, were entirely wrong. Miasma theory – that air filled with killer particles from decomposing matter caused disease – was broadly accepted in Victorian Britain and Chadwick was one of its greatest proponents.

As he approached his 90th birthday and, as it turns out, his death in 1890, Chadwick formed the Pure Air Company. This gave rise to the exciting prospect of a capital blighted by smoke-belching industrial chimneys being blessed with what was termed 'London's Eiffel Towers'. Like Edward Watkin and

Public health reformer Edwin Chadwick. In the nineteenth century he proposed construction of huge towers in London to suck cleansing, fresh air down to street level.

everyone else planning a tall structure at the time, people were keen to associate with the Parisian success. But rather than a mere trifle of entertainment or place to shop, Chadwick's towers would serve a much-needed social benefit; delivering air from on high and distributing it at street level.

Announcing his new enterprise to a Royal Society of Arts symposium that was chewing over issues around sewage disposal, one of his specialist subjects, Chadwick explained that miasma could be beaten if sufficiently tall towers were constructed. They would exceed the heights of neighbouring industrial smoke stacks and even the Eiffel Tower itself. *The Builder* magazine was impressed, if bored. 'Sir Edwin concluded his somewhat prolix communication,' it reported, 'by advocating the bringing down of fresh air from a height, by means of such structures as the Eiffel Tower and distributing it, warmed and fresh, in our buildings.' The towers would 'draw down air, by machinery, from the upper couches or strata of air and distribute it through great cities, like the Metropolis'.

Fresh oxygen could even go directly into slum houses, benefiting many families barely surviving in squalid, unsanitary conditions. Industrialisation had led to great numbers of people living in close proximity, sharing the same air. Many parents not only slept in the same rooms as their children, but often with other families and frequently with, or near, farm animals. Rotten food, rotting corpses, dead animals and sewage were rarely far away. People even kept what they euphemistically called 'night soil' that they could sell later for fertilizer. In 1861 the Statistical Society of London visited a single room occupied by five families, four of which ate, sat and slept in a corner each, with the fifth family in the middle. One woman told investigators: 'We did very well until the gentleman in the middle took a lodger.'

In 1868, London suffered its Great Stink. It was hardly unexpected. From the 1840s to 1860s, the smell of the capital was overwhelming. Across the country outbreaks of cholera killed tens of thousands of city dwellers. Miasma was blamed. But if Chadwick's plan to suck clean air from the sky to the ground had logic, it was the wrong logic. Against a growing body of evidence – although in line with prevailing opinion – he ignored the criticality of water cleanliness. Instead, *The Times* reported, Chadwick called for the 'complete drainage and purification of the dwelling house, next of the street and lastly of the river', in effect taking away waste from homes and adding it to the water.

This, though, was the parody of Chadwick's position. He had been a powerful advocate for urban drainage and sewerage systems, and although his contemporary Joseph Bazalgette is seen as London sewers' hero, Chadwick had also campaigned against the degrading and dirty conditions caused by industrialisation. Nevertheless he stuck to his fundamental position on miasma: bad air caused disease. He had high-profile support too. Florence Nightingale, by now famous for her nursing work during the Crimean War, was also a miasma-theory advocate; believing that measles, smallpox and scarlet fever were the fault of the new habit of building houses with drains below through which odours could escape and infect the masses. Nightingale proved a much more amenable character than Chadwick and from 1870 onwards had the ear of more people when she talked about sewage.

However, by failing to make the connection between clean water and disease, even though Dr John Snow had discovered this link at about the same time, Chadwick and Nightingale, along with most of the rest of the scientific community, were focusing on the wrong cause. Even though Snow's theory was gaining acceptance following a cholera outbreak in an area not covered by London's newly installed sewage system, Chadwick continued to be an exponent of the importance of clean air.

It took many years for miasma theory to be discredited, but even that should not lessen the contribution that both Chadwick and Nightingale made in emphasising the need for clean air. Even as late as the 1950s, poor air quality was dispatching Britain's urban poor in unacceptable numbers. 'It is time that the air we breathe is recognised to be as important as the water we drink,' thundered the *British Medical Journal*. 'We still treat the air over our large towns as a sewer; into it chimneys belch thousands of tons of harmful grits, solids and gases each year. Thus we have failed to learn the lesson of the Thames.'

Chadwick didn't live to see his miasma-busting towers. Nor has anyone else. They had no realistic possibility of gaining acceptance, largely because of Chadwick's age when he conceived them – 90 isn't the ideal time to start building Eiffel Tower-sized structures – but also because of his almost complete lack of interpersonal skills. Universally despised by both rich and poor, he was, according to one biographer: 'The most unpopular single individual in the whole United Kingdom.' As the architect of the workhouse, he was hardly the labouring classes' greatest hero. To make matters worse, he also wanted to restrict the sale of alcohol, one of the few legal substances that made many lives worth living. At the other end of the social scale, colleagues on committees or in his clubs found him quick to express forthright opinions in derogatory terms.

He had the character of 'the bore, the fanatic and the prig', said one. The passing of the years did little to mellow Chadwick, according to his friend Sir John J. Macdonnell: 'It is one of the few unquestioned privileges of old age to be a bore, and this great man had, I fear, discounted too early, too freely, and too heavily this privilege. One reason was that he babbled too much, not of green fields, but of sewage.'

In the end, Chadwick's Pure Air Company built no towers. Another of his schemes, to train rider-less fire horses to gallop automatically to burning buildings whenever they heard a fire alarm, didn't get out of the starting blocks. And to the dismay of generations of children since, he also failed to have spelling tests abolished in schools. They were 'quite unnecessary', he said, particularly as spelling lessons cost two-thirds of the sum spent on elementary education. But, as befits the first president of the Chartered Institute of Environmental Heath, Chadwick did want the Pure Air Company to survive and to provide a social benefit. Reflecting the teachings of his friend and benefactor Jeremy Bentham (whom we will meet next), any profit, Chadwick said, would not be large in order that the greatest good could be done for the greatest number of people. He was right in that the Pure Air Company wouldn't make a large profit. In fact, it made none at all, and no pure-air towers either. As he took his last breath of clean, fresh air – he lived in a nice part of town – his exciting project remained unfulfilled.

43

BENTHAM'S ALL-SEEING PANOPTICON

It's Big Brother's Big Brother. More than 150 years before George Orwell, philosopher Jeremy Bentham designed the panopticon; a building that would be 'all-seeing' and where people, fearing observation even when no one was present, would feel compelled to behave. Simply designed, ruthlessly efficient and although almost wholly ignored, not least by King George III who pulled funding from the first one, it was said that panopticon buildings would do good for society, good for the people being observed and good for the public purse.

Reflecting Bentham's philosophy, the essence was this: panopticon buildings made people behave through their simple but practical design: a circular or semi-circular block around a central base, generally a tower, from which surveillance could be conducted – or not. Every individual cell, indeed every individual, could be under watch at any time of night and day. Quite possibly – and here is the panopticon's genius – in reality no one would be monitoring very much at all. Everyone who was forced to reside within the building for whatever reason would see the tower, but not be able to penetrate its windows to see what lay beyond. Warders may be there, keeping watch, ready to enforce order, happy to punish. But equally they may be absent, and it is this concept of the uncertain, that

Big Brother *could* be watching you, rather than definitely is, that gave the panopticon its power of control.

Children, at desks in isolated cells, would behave exceptionally and attend to their studies dutifully in panopticon schools, where an omnipresent teacher could descend on them whenever attention wandered. Prisoners would conform to their gaolers' rules when not at work on the treadmill or spinning looms, which Bentham, one of the eras great liberal thinkers, believed would be productive activities. The insane would expect to be under constant watch in their panopticon hospital or asylum. Any attempt to flee would be madness. For the panopticon was essentially a building of mind control. Bentham called it 'the inspection house ... a method of controlling mind over mind'. While surveillance was a possibility rather than a certainty, pupils, prisoners or patients would indulge in self-surveillance. The belief that one is under constant watch can do much for the mind.

Reformer Jeremy Bentham's cadaver (with wax head) at University College London.

Bentham's panopticon manifesto was written in quaint, polite terms. 'The building is circular,' he explained. 'The apartments of the prisoners occupy the circumference. You may call them, if you please, the *cells*. The apartment of the inspector occupies the centre; you may call it, if you please, the *inspector's lodge*.' In their cells, inmates could see little but the tower. Even shadows out were designed to prevent inspectors revealing their presence. Each cell ran for the width of the building with an inside window looking out on the tower, and an outside one allowing light to pass through. Instead of doors, simple openings spaced so no one inmate could see another eradicated both noise and the chance of light from a half-opened door betraying the presence of a warder or teacher. 'All play, all chattering – in short, all distraction of every kind – is effectually banished by the central and covered situation of the master, seconded by partitions or screens between the scholars, as slight as you please,' wrote Bentham. 'That species of fraud at Westminster called *cribbing*, a vice thought hitherto congenial to schools, will never creep in here.'

By design, through fear, the panopticon would 'introduce tyranny into the abodes of innocence and youth,' which, although not easily reconciling with his liberal tendencies, Bentham thought a jolly good thing. Conversely, the panopticon prison would put an end to random acts of cruelty by warders. There was a financial advantage too. Because inmates and schoolboys would control their own behaviour, panopticons required far fewer staff and thus the wage bill could be slashed. Bentham's death in 1832 was just two years before new Poor Laws heralded the construction of workhouses, when millions, destitute and desperate, threw themselves on the mercy of their parish. In return for food, accommodation and ritual humiliation and cruelty, the state would expect absolute control on every aspect of their lives. The panopticon workhouse would be perfect. Many nineteenth century institutions – workhouses, prisons, hospitals – owed much to Bentham's design, if not his philosophy, which was to provide: 'the greatest happiness for the greatest number of people'. In many respects, he fell short on this one.

While plenty of buildings owed a nod to panopticon design, in reality almost no truly authentic such buildings were constructed, certainly not in Bentham's lifetime. This was unfortunate – and costly too. Although very wealthy, Bentham spent a large sum of money developing the panopticon prison, eventually persuading the government to provide a location. His plans were scuppered when George III refused to authorise purchase of the site. In 1813, Bentham received £23,000 to go some way towards making good his loss. It was small recompense for his efforts. If only the king had dug deeper, institutions such as the workhouse might have been creepier, but somewhat more humane.

Despite his losses on the panopticon prisons, Bentham left a sizable legacy to the University of London – what today is University College, London – and to Edwin Chadwick, the creator of the Pure Air Company. Ending his life an absolute rationalist, on his own request and to show that death really is the end and resurrection a myth, his body was dissected, embalmed, re-dressed in his own clothes – large hat, cutaway coat, nankeen trousers – and propped in a chair where it can be seen to this day in a display case at the university's Gower Street base. Bentham's head, sadly, is not his own, after the embalming process went wrong, leaving his face disfigured. Instead he now has a wax one. The real head, having survived safely in the college for nearly 150 years, was stolen by students in 1975 and held for ransom. After going missing several times subsequently, it is now kept in storage at an unknown location. His legacy, though, lives on; so to that extent, for Bentham at least, there is an afterlife. Benthanism gained a significant philosophical following that was both radically liberal and ultra-controlling ('bossy Benthamites', historian A.N. Wilson calls them).

Today panopticism is not much more than a concept in criminology and social science; brought back into fashion by French philosopher Michel Foucault whose 1975 work *Discipline & Punish: The Birth of the Prison* is very widely read in very small circles. 'The major effect of the Panopticon [is] to induce in the inmate a state of conscious and permanent visibility that assures the automatic functioning of power,' writes Foucault. 'Bentham laid down the principle that power should be visible and unverifiable.' And the concept is retained in cities and towns throughout the world through the use of CCTV cameras. We can all be seen – possibly – whatever we do, wherever we go. Panopticism survives, even if the buildings never made it.

44

THE SELF-CLEANING HOUSE

Frances Gabe hated housework. 'A thankless, unending job,' she called it. But if for much of her lifetime in the twentieth century a woman's place was in the home, she was going to create a home she didn't have to clean. Inspired, thought many. Completely nuts said others. But Mrs Gabe, coming out from a divorce and looking for something to occupy her (other than the housework) did build the world's first, and only, self-cleaning house.

If, as the Vatican decided in 2009, the washing machine did more to liberate women in the twentieth century than the pill or the right to work, Mrs Gabe was going to go one better. With sixty-eight mechanisms that take the effort out of cleaning, the whole 30ft x 45ft house at Newberg, Oregon that she built from scratch in her 60s is one giant washing machine. At a push of a button, small ceiling-mounted devices in every room would spring into action, running through an entire cleaning-drying-heating-cooling cycle and reaching deep into every nook and cranny with no need for anyone to deploy a duster or manhandle a vacuum cleaner. In a passable impression of an industrial car wash, jets of soapy water fired into action first, followed by a quick rinse and finally a heated blow dry. The full cycle took forty-five minutes. To ensure no damp patches remained, the floors of the house sloped slightly so that excess water drained away.

The thirty-eight patents for the house and its devices included bookshelves that dusted themselves and contained books with self-cleaning jackets; a fireplace that removed its own ashes; and a bathroom suite comprising of sink, shower, lavatory and bath that squirted themselves with detergent and had a good scrub after ablutions were complete. In the kitchen, the standard dishwasher was replaced by integral cleaning mechanisms in the cabinets, so that dirty crockery could be stacked in its rightful place and washed ready for the next meal. And the washing machine was redundant, as clothes were laundered as they hung in the wardrobe. You simply took off your dirty clothes, hung them up and set the cycle. An hour later they were clean and dry, ready to be worn afresh.

There were downsides. Fixtures, fittings and fabrics had to be completely waterproof (although none of Mrs Gabe's furniture was of Edison's concrete type). Carpets were entirely out of the question, as was wallpaper. 'Nothing but dirt-collectors,' said Mrs Gabe.

Conceived largely in the 1950s, when mod cons were coming into the home and people were beginning to afford new labour-saving devices, the house was lauded and laughed at in equal measure, with the Massachusetts Institute of Technology, one of the world's leading centres for innovation and enterprise, declaring Mrs Gabe one of America's leading female inventors. On the sitting room wall she displayed the original patents – some of the world's longest and most complex – safely encased in plastic, so guests could admire them as they relaxed on the plastic-wrapped upholstery.

Mrs Gabe, who lived in the original prototype of the house well into her 90s, thought that time spent cleaning which instead could be spent with one's family or on self-improvement, was time wasted. She spent twelve years working on the inventions in her garden, where she installed a series of shower stalls to investigate the behaviour of water at different pressures and on different substances. The house, she believed, would appeal to those who found the work/life balance difficult to manage and the elderly or those with disabilities who simply couldn't keep up the running of a home without help. Women, in particular, would admire the self-cleaning house – unless they are cleaners. 'The problem with houses is that they are designed by men,' Mrs Gabe explained. 'They put in far too much space and then you have to take care of it.'

Despite the attractiveness of the idea and the thoroughness with which Mrs Gabe thought of every possible defect, the concept failed to take off commercially and the patents expired. While the cleaning element may sound practical, living in a house bereft of soft furnishings, with slightly sloping floors and furniture that's encased in plastic, clearly isn't everyone's idea of a comfortable home. And things haven't always gone smoothly. Once the ceilings were damaged in a flood, then the house was hit by an earthquake, putting most of the self-cleaning mechanisms out of action. But Mrs Gabe lived a long and happy life at the house, becoming something of a minor celebrity. Even her oddities were celebrated. Door frames were designed deliberately low so that guests had to bow to her when they came in and a sign outside read: 'Please do not trample the poison ivy or feed the bull.'

THE JAW-DROPPING DIET

It is said that nearly 1,000 years ago William the Conqueror, worried about his weight, went on a self-imposed 'alcohol only' diet. And in the sixteenth century a Venetian man called Luigi Cornaro made something of a name for himself by living to 100 on a diet which appears not to have consisted of much more than eggs and wine. But diet regimes as the popular fads we know today, only really began to emerge in the nineteenth century. At first there was the Reverend Sylvester Graham, the father of Graham crackers, who promoted simple foods and vegetarianism. Then, in 1863, an undertaker called William Banting wrote what was really the first handbook for dieters, *A Letter on Corpulence*, which became a bestseller. It warned slimmers off starchy, sugary fare in favour of lean meat, eggs and vegetables and has been seen as a forerunner of the modern Atkins Diet.

Then, towards the final years of the century, the American Horace Fletcher arrived on the scene to cash in on the new dieting craze. Today few of us have heard of Fletcher and even fewer are following his eating plan which would see us chew each morsel of food over and over again. Yet in his day Fletcher's diet gained mass popularity and advocates included the likes of the oil tycoon John D. Rockefeller and the authors Henry James and Franz Kafka. His chew-chew diet would make him a fortune and earn him the nickname 'The Great Masticator'!

Fletcher was a successful art dealer but by the time he reached his 40s he wasn't happy. He was just 5ft 6in but weighed more than 15 stones and felt, he said later, like 'an old man'. Then, in the early 1890s, he was turned down for life insurance. It was this that led Fletcher to drastic measures. In 1898 he was killing time over a hotel meal on a business trip to Chicago when he got to thinking about eating styles and its effect on health. He began devising an extreme food plan to battle his own bulge. And within just five months of trying it out he had lost more than 60lbs.

So how did it work? Fletcher was inspired by the advice of former British prime minister William Gladstone to his children, to chew their food 32 times, once for each tooth. He made this notion the nub of his new diet. Food, he said, needed to be properly masticated before it was swallowed. In order to achieve this, dieters should munch whatever they were eating to the point where it became completely liquidised. The precise number of chews would depend on the type of food. Bread, for instance, might take as much as seventy chews. A shallot might need as much as 700. The idea was that all the effort of this would slow people down, so that they consumed less in total. If a food couldn't be entirely chewed down into liquid form? Simple – spit it out. As well as stopping you eat too much, Fletcher also thought that his chewing methods would allow the body to absorb the most potential goodness from the food.

Fletcher's first chewing work emerged in 1898 and he went on to publish several works over the next twenty years outlining his philosophy. In January 1904 the esteemed medical journal *The Lancet* joined a chorus of approval reporting that: 'a more generally beneficial doctrine could hardly be chosen.' Fletcherising, as the new chewing diet was dubbed, became all the rage on both sides of the Atlantic. Muncheon parties even became popular where food would be served and people would chew each bite simultaneously for five minutes before a bell was rung indicating that they could swallow. Fletcher became good friends with J.H. Kellogg, the inventor of cornflakes, who recommended his methods to clients. Even a section of the British army road tested his methods, though with mixed results.

Most diets, like Fletchers, are fads. They come and go as quickly as the fast food that so often makes them necessary in the first place. And when Fletcher died in 1919, from bronchitis, his regime was already losing favour as other diet plans appeared. Yet even modern experts believe there's some method in Fletcher's practices. Several scientific studies now show that the longer people take over their dinner the less likely they are to overeat. And along with all that chewing, Horace had some good, if basic, practical advice, such as only eating when you feel hungry. When the historian Sir Roy Strong tested Fletcher's plan and other diets from history he found that, though revolting, Fletcherising was one of the most effective historical diets in achieving weight loss in his subjects.

Of course Fletcher wasn't a doctor and his methods were largely based on his own personal eating plan. His assertion that Fletcherising your food would actually increase your strength and even help cure ailments like toothache look far-fetched today, especially as he seems not to have properly understood how digestion really worked or the benefits of vitamins and minerals. But Fletcher clearly felt personally invigorated by the chewing practice. Aged 58 he supposedly beat a group of young Yale college athletes in a series of tests at the gymnasium there.

Ultimately, however, his way of eating didn't catch on permanently. Perhaps Fletcher was simply more patient than the general populace. Most people, it turned out, couldn't be bothered to spend all that time chewing. His eating methods certainly might not be all that appealing, especially if you had to share dinner with someone who followed them and were expecting some lively conversation. But they were probably a lot more effective than William the Conqueror's. The king's booze-fuelled weight-loss plan died along with him when in 1087, and still fat, he was mortally wounded after falling off his horse.

46

A NUTTY PLAN TO FEED THE MASSES

One writer has described it as 'post-war Britain's equivalent of the Millennium Dome.' Another called it a 'complete fiasco'. In 1945 Britain and its empire were in the grip of post-war austerity and suffering from severe food shortages. Most of all, the nation needed fat, and lots of it. From July that year, Clement Attlee's energetic new Labour administration held the reins of power and its minister of food, John Strachey, helped cook up a grandiose plan to solve the problem. Aimed at ensuring a steady flow of cheap vegetable oil, he aimed to get the nation back on its feet. The answer, he believed, was nuts!

One of his first moves in office once he became minister of food in 1946 was to introduce bread rationing, which hadn't even been rationed in the war. 'Starve with Strachey' was a popular catchphrase that year. So when the great groundnut plan came along, he leapt on it. It was a scheme that was to haunt him and his government for the rest of the decade and to end in rather comical catastrophe.

The original idea had come from Frank Samuel, managing director of the United Africa Company. Groundnuts, another name for peanuts, were an excellent source of oil and already grown on a small scale by locals in East Africa. But the region, most of which was still part of the British Empire, had vast tracts of uncultivated land. All it needed, advised Samuel, was to plant this fast-growing crop and, using modern farming techniques, reap the reward.

The resulting fats could be used to feed the British at home and bring wealth to Africa too. He thought it too big for a private firm to do and presented his findings to the government who commissioned a report from John Wakefield, former director of agriculture in Tanganyika, today's Tanzania. Wakefield's somewhat over-optimistic report recommended clearing 3.2 million acres for cultivation in the region. The sensible thing might have been to undertake a pilot scheme, as a junior minister suggested. But Strachey was in a hurry. Instead he gave the colossal project a green light, authorising £25 million to cultivate 150,000 acres of scrubland a year in Tanganyika, along with construction of a new port, railway and roads. The yield would be 600,000 tons of groundnuts annually, helping to boost the local economy and put a big dent in Britain's food bill.

Almost nothing went right from the start. As Alan Wood, a journalist who wrote the definitive history of the affair back in the 1950s put it: 'they were proposing a colossal engineering and agricultural revolution, something comparable on a small scale to the Soviet Five-Year Plans, without even realising what they were doing.' No proper studies were undertaken on levels of rainfall in the area, essential for a crop that needed plenty of water. No thorough survey of the land or soil had been done or analysis of crop yields. These only started once the scheme was already well under way.

Work began in 1947 in the Kongwa region of the country. Yet there were not enough tractors to do the work, so the project's leaders had to scour the globe for second hand ones, including US army surplus from the Philippines. Another problem was the local railway. The single-track line didn't have enough capacity to transport all the men and materials up from the port at Dar-es-Salaam into the interior. Then it was washed away in a flood. Once the few tractors that did make it arrived, it transpired that they weren't up to the job of clearing the land, a much harder task than originally envisaged. One report suggests that three quarters of the machinery broke down.

When the team did find a way of clearing the trees that involved using three bulldozers and a chain, they sent an order for ship's anchor chains to help. London cancelled the request thinking it was a joke. A vast army of thousands of men were engaged in the work. But they repeatedly went on strike. To top it all the workers were plagued by scorpions and swarms of bees that left bulldozer operators hospitalised. Agriculturally things weren't going much better. Once the groundnuts had been sown periods of drought turned the clay soil to concrete. By the end of the first year just 7,500 acres of groundnuts had been planted.

In spring 1948 new management was brought in under a new body, the Overseas Food Corporation, with a Major-General Desmond Harrison in charge on site. He tried to run the scheme as a military operation, but he soon had to return home sick. Towards the end of the project, which was finally abandoned in 1951, desperate officials tried growing sunflowers instead. There turned out to be too much sun and the crop failed. It was left to a new Minister of Food in a new government, Maurice Webb, to share the bad news about the scheme. One telling sentence in a government document of the time admitted: 'The groundnut is not a plant which lends itself readily to mass methods over vast acres.' It was also estimated that the crops had cost six times more to produce than they were worth.

In the years that followed the cry of 'groundnuts' was enough to reduce the House of Commons to giggles and the scheme became synonymous with the failure of big government projects. Not so funny was the bill for the whole debacle, which came in it at around £49 million, estimated at more than £1 billion in today's money. Strachey's reputation never recovered.

For a government lauded for achievements such as setting up the NHS, the great groundnuts cock-up is an episode the administration's fans would rather forget. Decades later

Labour leader and thinker Michael Foot – a Labour MP in the late 1940s – bemoaned the scheme's failure saying it could have been a blueprint for feeding Africa. He said: 'I think they could have gone ahead with many other schemes of that nature – both from the point of view of trying to assist the countries in Africa as well as to help the food policy here. Because that scheme failed they rather got cold feet, but I think that was a pity.'

47

RAVENSCAR: THE HOLIDAY RESORT THAT NEVER WAS

The place 'has been happily named, offers a rare combination of nearly all the natural advantages required in a health resort, watering place, and holidaymakers' home'. So read the brochure for a grand, new seaside town, one which aimed to trip off the tongue as easily as Blackpool or Brighton. It was to be another of the places to which smog-bound urban Victorians could easily flee. The florid description of Ravenscar certainly made it seem like the ideal venue for a relaxing break away from the stresses and strains of city life. Yet there was something strange about the tempting pleasures outlined in the brochure. For most of the luxurious resort of Ravenscar didn't yet exist.

The Victorians have been credited with inventing the idea of the seaside-holiday resort. Indeed most of the traditions associated with the British seaside such as the promenades, piers and donkey rides emerged during the nineteenth century. Small towns and villages were transformed into bustling popular destinations as people escaped the sooty cities created by the industrial revolution to enjoy the sea air and cheap entertainment. These new pleasure centres also became more accessible thanks to the century's railway boom.

By the 1890s many resorts were well established and the appetite for the seaside showed no signs of dissipating. The list of places which had cashed in on this boom was long. As well as Brighton and Blackpool was Skegness, Southport, Bridlington in the north and Bournemouth and Weymouth in the south – to name just a few. In the 1890s developers drew up plans for a new destination in Yorkshire designed to rival Scarborough, which had become another great seaside Mecca. The hamlet of Peak, where the new town was to be constructed, was a stunning location. Set 600ft up on a dramatic headland its location looked out over the beautiful Robin Hood Bay and behind it lay the wild expanses of the North Yorkshire Moors. Crucially, in 1885, the spot got a train station when the Scarborough and Whitby Railway was pushed through the hilly landscape, making access for tourists straightforward.

The line had been built thanks to a William Hammond, who owned the local castellated pile, Peak Hall. As chairman of the North Eastern Railways Company, he had been instrumental in the building of the new station near to his home. But perhaps the fact that he insisted on getting an extra tunnel built so that he didn't have to see the line was an ill omen for any future development of the rural idyll.

In 1895, following the death of Hammond, the couple's four daughters sold the hall and its land. The hall subsequently opened as the Raven Hall Hotel and a golf course was built. By 1897 The Peak Estate Company, led by an entrepreneur called John Bland, had bought the land for £10,000 and planned to make the hotel the hub of their new resort. A brickworks was opened nearby to supply the anticipated building boom. The idea was to sell 1,500 plots of land for houses and 300 men were soon put to work building sewers, mains water supply and even laying out a grid of streets. There was Roman Road, Saxon Road and some even got grand names like Marine Esplanade. As well as shops and tearooms there were to be hanging gardens too.

The name of the town was to be changed to Ravenscar after the Viking raiders who had once plundered the Yorkshire coast. The Vikings had used an emblem of a raven on their banners, while 'scar' was a Norse word for cliff. In a move that

A vintage railway poster advertising the charms of Ravenscar, the phantom Yorkshire seaside resort.

was certainly premature, you could even get a guidebook to the place. There appeared to be no question that Ravenscar would attract a multitude of visitors and investors eager to make the most of this spectacular addition to the nation's seaside attractions. The company's brochure continued: 'With the huge population of the West Riding and Midlands behind it, annually overcrowding the existing outlets to the sea, it needs little prescience to foresee that the future of Ravenscar as a watering place is practically assured.'

However, when prospective buyers for the properties arrived most were disappointed. What there was of a beach below the cliffs was rocky and Ravenscar's altitude meant that the shoreline was pretty inaccessible anyway. Despite the rather dubious claim that in 1902 it had enjoyed seventy fewer rainy days than Scarborough, the climate could be very inhospitable. The cliffs were lashed by high winds and sea mists would often settle over the high ground. Prospective buyers travelling to inspect Ravenscar were offered a refund on their train fares if they bought a plot of land, but too few appreciated the charms of the place. In 1911, the company behind the scheme went bust.

Today the street layout is still evident. But where you'd expect to find rows of terraces dotted with B&Bs, there are isolated Victorian houses which look as if they have been plucked from the streets of more successful resorts like Margate or Morecambe. The population of Ravenscar today is just a few hundred; it's not the bustling town that was hoped for. In 1965 even Ravenscar's station closed, a victim of the swingeing Beeching cuts, based on the report by Dr Richard Beeching which saw many supposedly unprofitable local lines shut down. It was a far cry from the railway poster which had once proudly boasted of the charms which awaited visitors to the Yorkshire paradise. It read: 'twixt moors and sea … magnificent undercliff and hanging gardens … most bracing health resort on the East Coast'.

48

WHY SMELL-O-VISION STANK

First there had been the movies, then there had been the talk-
ies – now there would be the smellies! So went the hype for the
much-anticipated 1960 movie *Scent of Mystery*. The film's pro-
ducer Michael Todd Jr. believed that he was on the precipice
of a new era in movie making, one in which audiences would
not only be able to see the action but sniff out the plot too.
This would be achieved thanks to a revolutionary new process
which would enliven the action. Appropriate odours were to
be pumped into the cinema as the movie's storyline developed.

The introduction of Smell-O-Vision, as this new technology was called, may now seem a potty episode in the history of
cinema. But in the 1950s and 1960s studio bosses were worried
about the growing lure of television and desperately trying to
find ways to keep the cinemas full. Of the array of new technical ideas which aimed to enhance the big screen experience,
Smell-O-Vision was the most intriguing. The science behind
it was provided by a Swiss inventor called Hans Laube. He'd
already come up with a successful process of getting rid of
smells in cinemas when it occurred to him that they could also
be added back in too.

He wasn't the first to try stimulating the olfactory senses in
movie auditoriums. As far back as 1916, a newsreel about an

American football game from LA's Rose Bowl had been com-
plimented by one cinema in Pennsylvania using rose oil and a
fan. A Boston cinema sprayed lilac oil through its ventilation systems over the opening credits of the 1929 film *Lilac Time*. Laube,
however, was the first person to propose a precise system that
delivered a whole range of odours to order. In 1939 he travelled to the New York Fair to demonstrate what he called his
'smell-brain'. He made a short film to show how smells could
be delivered to individual cinema seats. The idea aroused some
interest in the papers, but studio bosses remained unexcited.

Then, in the late 1950s, Todd Junior strode on to the stage.
He was the son of famous movie mogul Mike Todd Senior
and had worked with him on the wrap-around, big-screen
phenomenon called Cinerama which used three projectors to
produce a dramatic image. He'd also helped with his father's
1956 movie *Around The World In 80 Days*, which became a
smash hit and won an Oscar. After Todd Sr. died in a plane crash
in 1958, Todd Jr. took up the reins of the business. He wanted
to go further in the field of movie innovation. Having already
discussed the idea of adding smells to films with his dad, Todd
decided to employ Laube, paying for his experiments with a
'smellies' system at Chicago's Cinestage cinema. There was one
proviso – what Laube had been calling Scentovision was to be
changed to the catchier Smell-O-Vision.

Laube began refining his system. Smells were to be pumped
to cinema-goers through tubes beneath their chairs using
a series of vials filled with the different fragrances, which
would be triggered by the film's soundtrack. Each vial contained 400cc of smell, enough for 180 performances. A mile
of tubing in each theatre would get the aromas to the noses
of the paying public. The process was to be debuted as part of
the comedy thriller *Scent of Mystery*, which was to be shot with
Smell-O-Vision in mind. The film starred Denholm Elliott as
a holidaymaker in Spain who uncovers a conspiracy to murder
an American heiress, played by Elizabeth Taylor.

Thirty different odours would be emitted during the
showing of the film. They ranged from garlic to tobacco and

bananas as well as bread and shoe polish. When, in one scene, wine casks are smashed a grape fragrance was to be released. To add spice, the smells were worked into the plot too, providing clues to help Elliot's character get to the bottom of the mystery. Elizabeth Taylor, for example, is identified by her expensive perfume, the killer by the smell of his pipe. In the run up to the movie's release the press went wild for the idea. Todd helped whip up the frenzy with pun-filled publicity. 'I hope it's the kind of picture they call a scentsation!' he said. Once the film had debuted in three specially modified cinemas Todd had plans to speedily roll out the new technology to 100 more.

Heightening the tension, it transpired that Smell-O-Vision would be going head to head with a rival system – AromaRama. This had been introduced by Charles Weiss and Walter Reade Jr. for a documentary film *Behind The Great Wall* which hit cinemas in December 1959 just weeks before *Scent of Mystery* was released in February 1960. Weiss and Reade's smell mechanism delivered fifty-two odours such as jasmine into the cinema via the air conditioning system. The contest between the two was dubbed the 'battle of the smellies' by *Variety* magazine. In reality *Scent of Mystery* was a much bigger deal. Its system was more elaborate and cost three times as much to install. At its release in New York, Los Angles and Chicago, *Scent of Mystery* was even preceded by an animated short film about a dog which had lost its sense of smell, also made for Smell-O-Vision.

The film's reviews were mixed. *The Hollywood Reporter*, for instance, announced that: 'the story itself has a distinct charm, fascination and humorous edge'. The *New York Times* critic Bosley Crowther was more harsh, branding it an 'artless, loose-jointed "chase" picture set against some of the scenic beauties of Spain'. More disastrously, the Smell-O-vision process was not, as Todd had hoped, described as a 'scentsation'. Audiences complained of strange hissing noises. Some were irritated that that the cinema's ventilation systems couldn't get rid of one smell before another was released. Others had difficulty picking up the odours which apparently led to the sound of heavy sniffing pervading the cinema. *Time Magazine* summed it up like this: 'Most customers will probably agree that the smell they liked best was the one that they got during intermission: fresh air.'

Laube hastily made adjustments, but it was too late. *The Scent of Mystery* wasn't making headlines at the box office and the furore about the technology was fading. Some time later the movie was re-released with the much more humdrum title *Holiday In Spain* and without the smells, which had the effect of leaving audiences even more bemused. The *Daily Telegraph* reported: 'the film acquired a baffling, almost surreal quality, since there was no reason why, for example, a loaf of bread should be lifted from the oven and thrust into the camera for what seemed to be an unconscionably long time'.

In the coming years Todd gradually faded out of the movie business, moving to Ireland. Laube quietly disappeared from the scene, while the AromaRama system sank too. There were attempts in the ensuing decades to revive the idea. In the 1980s an Odorama version of the film *Polyester* was released with scratch and sniff cards. To date, however, *The Scent of Mystery* remains the only movie to have been shot in Smell-O-Vision and it seems unlikely that the technology will be wafting into multiplexes anytime soon.

Japanese scientists have recently been working on a gadget that can accompany TV shows. But if a 1965 prank is anything to go by they might be wasting their time – the mere power of suggestion may be enough. That year the BBC played an April Fool's joke where it aired a spoof interview with a man who said he had invented a type of smell-o-vision for television, and then demonstrated by chopping some onions and brewing a pot of coffee. Duped viewers rang in to say that they had, indeed, smelled those very aromas coming through their TV sets.

49

THE METAL CRICKET BAT

The soothing sound of leather on willow. It is one of the joys of the sporting summer. The sound of leather on metal? It doesn't, somehow, have the same ring to it. Fiery Australian cricketing legend Dennis Lillee thought differently. When he walked out to bat at a sun-drenched test match between England and Australia on 15 December 1979, he was wielding a revolutionary blade which would lead to one of the biggest controversies and on-pitch bust ups the sport had seen.

Lillee, one of the game's brightest stars and tempestuous characters, had taken to the field at Perth's WACA stadium clutching a bat made out of aluminium. At the start of its second day, the match was delicately poised with Australia in some trouble on 232 for eight. Lillee was a superb fast bowler who took 355 test wickets during his career. He was not, however, known as a great batsman. So as he strode to the middle with his new metal wand, perhaps he hoped that this bit of kit would add a sprinkle of magic to his performance.

Lillee had ended the first day's play on a score of eleven not out, using an ordinary wooden bat. But on that second morning, he had chosen to take guard once more with the new invention. The ComBat, as it was called, had been created with the help of a businessman friend as a cheap alternative to the traditional wooden bats. Lillee had actually used the bat before in a test match, unleashing it at a game against the West Indies in Brisbane, where it had only raised some wry smiles. It certainly didn't make much of an impact on the contest. The ball had clunked against it only once, then thudded against his pad leaving him out, leg before wicket, for no runs. Now, just two weeks later, an unperturbed Lillee stood facing the equally charismatic and dangerous England bowler Ian Botham with the metal bat in his hands once again. On the fourth ball he received, Lillee saw his chance for a big shot, driving the ball towards the boundary, with England fielder David Gower in hot pursuit. As the ball left the bat, it made a distinctive, somewhat tinny sound.

England's captain Mike Brearley soon guessed what was up. Moments later he ran over to the umpires Max O'Connell and Don Weser complaining that Lillee's bat was damaging the ball. A heated discussion ensued with the umpires telling Lillee that he had to change the bat. Lillee stood his ground. Meanwhile, even Lillee's own captain, Greg Chappell, wasn't sure the bat was actually helping the Aussie cause. He thought the shot Lillee had hit, which went for three runs, would have gone for four if it had been hit with a normal willow bat and promptly sent his twelfth man, Rodney Hogg, out on to the pitch to deliver one to Lillee and retrieve the offending metal one. Still Lillee refused to budge.

For ten minutes debate raged in front of a bemused crowd, before Chappell himself intervened, marching on to the pitch with a wooden bat and ordering his team mate to use it. With his own skipper now against him, Lillee was persuaded to give in, but not before he'd hurled his metal bat high into the air in disgust, 40 yards back in the direction of the dressing room.

Lillee went on to make eighteen runs, caught out off Botham's bowling, with the side all out for 244. Ironically Australia went on to win the game by 138 runs and won the three match series 3-0. Lillee may have had good reason to be disgruntled. Using an aluminium bat wasn't actually against the game's rules. Law 6 stipulates that the bat is 38in in length, and no more than 4¼in wide. At the time it said nothing about what the bat was made from. Lillee might have pointed out that after equipment used in the world's major sports has always evolved. The first cricket bats were shaped like hockey sticks. Footballs started off as animal bladders.

The spat over Lillee's bat actually turned out to be something of a brief marketing triumph, stoking not only controversy but sales of metal bats too. At the end of the match, Lillee had got all the players to sign the bat in question. Brearley wryly wrote on it: 'Good luck with the sales'. Lillee has since admitted that the ComBats were based on aluminium baseball bats and really designed: 'not for Test cricket but for practice, for clubs, for schools, all that sort of stuff'. Cricket traditionalists took a dim view of the whole incident. The game was only just recovering from the turmoil over Kerry Packer's controversial World Series Cricket. The row over the breakaway competition had briefly threatened to split the whole sport apart when Packer created a competition just for his Australian TV channels and lured players away from their national cricket associations.

Wisden Cricketers' Almanack, that guardian of the game's morals, said that the aluminium bat 'incident served only to blacken Lillee's reputation and damage the image of the game as well as, eventually, the Australian authorities because of their reluctance to take effective disciplinary action'. But the international game's unamused chiefs did take action. In 1980 cricket's governing body changed Law 6 to state clearly that the bat should be made out of wood.

Metal bats do still exist, sometimes used in amateur forms of the game, but they are banned at test level. There were echoes of the Lillee debacle in 2004, when another Australian, Ricky Ponting, started using a wooden bat that featured a graphite strip. This was subsequently banned too. Millions of cricketing traditionalists are thankful that Lillee's innovation never caught on. Nevertheless, the remaining ComBats are now sought after by collectors. Lillee has since suggested that throwing the bat in a fit of temper might have been a bit much. Shortly after releasing his autobiography, *Menace*, he said: 'I cringe a bit at it now.'

50

BICYCLE POLO & OTHER LOST OLYMPIC SPORTS

The summer Olympic Games of 1908 was a strange affair. Hastily held in London when Rome was forced to pull out for financial reasons following the eruption of Mount Vesuvius, it is probably best known for its marathon race which turned into something of a farce. The winner, Dorando Pietri, collapsed as he entered the White City Stadium on the last leg of the race and was helped over the line by officials. The Italian runner was later disqualified.

In many ways the modern version of the Olympic Games, reintroduced in Athens in 1896, was still finding its feet. Both the 1900 and 1904 tournaments had been held as part of World Fairs, rather than as events in themselves. And in those early days of the ultimate sporting spectacle some unusual disciplines were included in the programme. At the 1900 Games, held in Paris, the tug of war was a big draw – in which, rather oddly, a mixture of Danes and Swedes teamed up against the French to scoop the gold medal. The Games also hosted an array of 'demonstration sports' vying for their official inclusion which included cannon shooting, fire fighting and even ballooning.

In that year the French had also been the only country to take part in the official croquet competition, taking all the medals. Four years later, the Games were being held on US soil, in St Louis, Missouri for the first time. Rather than try and beat the French at their ancient pastime, the Americans decided to invent their own version of croquet. They simply removed the letters c and t from the beginning and end of the word croquet to come up with roque. Though they tinkered with the rules a little, the crucial difference was that roque, unlike croquet, was played on a hard surface rather than grass. Only four players took part in the roque competition – all American. In the following years the sport briefly caught the public's imagination; by 1920 there was a national league. But as the century wore on interest gradually fizzled out.

The 1908 Games in London would see more intrigue. Its tug-of-war event almost turned into a riot when the US squad accused one of three British teams, made up of Liverpudlian policemen, of wearing spiked boots. The protest was overturned and the three different British teams went on to win gold, silver and bronze. Among the new sports was running deer, a shooting event which used targets rather than actual animals. Another sport that featured in 1908 was bicycle polo. This brand new game was invented by an Irishman and a retired champion cyclist Richard J. Mecredy. In 1891 he came up with a new spin on an old sport. 'Posh polo' had been around for centuries, though equestrian polo's official rules had only been drawn up in 1874. Why not, thought Mecredy, combine the drama of polo with the popular passion for the bike and create a 'common man's version of the sport of kings'?

As editor of a magazine called the *Irish Cyclist*, Mecredy published rules for the game in October 1891, following the first ever match between two Irish teams: Rathclaren Rovers and Ohne Hast Cycling Club. Early photos show players armed with mallets but no protective headgear. A contemporary report reassured the public that the game 'was not at all so dangerous as would appear from the title'. Over the course of the next decade bicycle polo caught on in Britain too, with clubs in Catford, Newcastle and Northampton and a national association formed. In 1901 the first international game was played in which the Irish soundly defeated England 10–5.

When the 1908 Games came along, bicycle polo's fans saw an opportunity to boost the sport's profile further, getting it accepted as a so-called demonstration sport, which allowed hopeful Olympic sports to promote their cause for regular inclusion. Ireland, which still dominated the world of bicycle polo, wasn't allowed to compete in most events that year as a separate nation as it was still part of the United Kingdom of Great Britain and Ireland. But with tensions over Irish independence running high, Ireland was allowed to field its own teams in hockey, polo, and in bicycle polo too. At the final of the bicycle polo at the Games in July 1908 an Irish team, which appears to have included Mecredy's own son, soundly beat a German side 3–1.

Bicycle polo did not feature at the fifth Olympiad in 1912 in Stockholm and, thanks to the First World War, the sport lost many of its finest players. However, by the 1930s, the popularity of bicycle polo was growing in Britain once again and regional leagues were introduced. By 1938 there were around 170 official teams across the country and 1,000 registered players. A guide from the time spells out the rules of how a match should be played. Two teams feature four mounted players on a marked-out field measuring 100 yards by 60 yards. Games, lasting ninety minutes, would see players in pursuit of the ball while skillfully manouvering their two-wheeled steeds. Each player wielded a wooden mallet with a head of 7in, aiming to knock the ball into the opposing goal. Foul play was definitely out – on no account should the mallet be shoved through an opposing player's spokes.

This heyday was short lived. Again war intervened and after the Second World War the sport faltered as a popular pastime, but just about clung on, especially in France. While it never again graced an Olympic stage, and it's pretty unusual to see a match being played at your local park today, an urban version of the game, played on hard courts, has found a new wave of popularity in recent years. There are still international matches, even for the grass version.

It is, of course, possible that one day bicycle polo might re-emerge at the Olympics. After all golf, not played at the

Olympics since 1900, is to return for the 2016 Games in Rio de Janeiro. There is one lost Olympic event that almost certainly won't be making a comeback – live pigeon shooting. It featured once, in 1900. To date it remains the only Olympic event from the modern Games in which live animals were killed. Almost 300 were shot. A Belgian, Leon de Lunden, won gold at the gory competition, felling a total of twenty-one birds.

ILLUSTRATIONS

We would like to express our thanks to everyone who has helped with the illustrations in this book. We have made every attempt to trace owners of copyright material and to attribute each correctly. We apologise for any omissions and will be pleased to incorporate missing acknowledgements in any future editions.

1 Pigeon-guided missiles
Image courtesy of the B.F. Skinner Foundation

3 The 'Spruce Goose'
Image courtesy of the Evergreen Aviation and Space Museum, www.sprucegoose.org

4 A sound plan for defence
Images copyright Peter Spurgeon, www.peterspurgeon.net

6 Tesla's earthquake machine
Original patent image from the US Patent Office

7 Edison's concrete furniture
US Department of Interior, National Park Service, Thomas Edison National Historical Park

11 London's Eiffel Tower
Images courtesy of Brent Archives, London Borough of Brent, www.brent.gov.uk

SOURCES AND RESOURCES

1 Pigeon-guided missiles
National Museum of American History Archives.
American Psychologist (Vol. 15, Issue 1, January 1960), pp. 28–37.
BF Skinner Foundation archive.
Psychology Today (14 July 2010).
BBC Online article (8 March 2000).
Daily Telegraph (7 Dec 2009).

2 The international 'hot air' airline
C. Fayette Taylor, 'A review of the evolution of aircraft piston engines',
    Smithsonian Annals of Flight (Vol. 1, No 4, 1971).
Clive Hart, *A Prehistory of flight* (University of California Press, 1985).
Russell Naughton, *Hargrave Aviation and Aeromodelling* (Pandora Archive,
    University of Canberra, 2007).
Chard Museum, Somerset holds many of John Stringfellow's documents,
    including the Henson and Stringfellow model planes.
London's Science Museum holds copies of the advertisements and other
    promotional material for the proposed international airline.

3 The 'Spruce Goose'
Evergreen Aviation Museum, www.sprucegoose.org
The Columbian (6 May 2003).
Popular Science (September 1945).
Popular Mechanics.
Air Power History (22 Sep 2007).
The Economist (12 Dec 1992).

Wood Based Panels International (1 June 2004).
The Press (1 Jan 2007).

4 A sound plan for defence

Richard Scarth, *Echoes From The Sky: A Story of Acoustic Defence* (Hythe Civic Society, 1999).
'Listening For The Enemy', *Cabinet Magazine* (fall/winter 2003).
Mark Denny, *Blip, Ping & Buzz* (The John Hopkins University Press, 2007).

5 The diabolical death ray

New Scientist (23 Dec 1976).
Jonathan Foster, *The Death Ray – The Secret Life of Harry Grindell Matthews* (Inventive Publishing, 2009).
Time Magazine (21 April 1924).
Fortean Times (October 2003).
South Wales Echo (30 July 2002).

6 Tesla's earthquake machine

Margaret Cheney, *Tesla: Man Out of Time* (Simon & Schuster, 2001).
David Hatcher Childress, (ed.), *Anti-gravity handbook* (Adventures Unlimited Press, 1994).
Earl Sparling, 'Nikola Tesla, At 79, Uses Earth To Transmit Signals: Expects To Have $100,000,000 Within Two Years', *New York World Telegram* (11 July 1935).
Nikola Tesla Museum, Belgrade, holds a number of Tesla's documents.

7 Edison's concrete furniture

R.W. Clark, *Edison: the man who made the future* (New York, 1977).
'Edison now making concrete furniture', *New York Times* (9 December 1911).
Michael Peterson, 'Thomas Edison's Concrete Houses', *American Heritage* (Vol. 11, Issue 3, winter 1996).

8 The misplaced Maginot Line

Time Magazine (18 Jan 1932).
Ian Ousby, *Occupation: The Ordeal of France* (Pimlico, 1997).
The Historian (22 June 2008).

9 The great 'Panjandrum'

Brian Johnson, *The Secret War* (Arrow Press, 1978).

Di James Kiras, *Special operations and strategy: from World War II to the War on Terrorism* (Routledge, 2006).

Atlantic Wall Linear Museum, www.atlanticwall.polimi.it

10 The first Channel Tunnel
Hansard (HL Deb 24 January 1929, Vol. 72), pp. 788–95.
Contemporary Review (1 June 1995).
The Independent (6 May 1994).
History Today (1 Nov 1996).
New Scientist (4 Feb 1971).
The Railway News (13 Dec 1873, 23 April 1883).
Telegraph (6 August 2010).
The Independent (2 April 2007).

11 London's Eiffel Tower
Original publicity materials in the Brent Borough Council Archive.
Country Life (19 May 1955).
Daily Graphic (14 April 1894).
The Guardian (14 March 2006).

12 Nelson's pyramid
Felix Barker and Ralph Hyde, *London as it might have been* (John Murray, 1982).

G.H. Gator and F.R. Holmes, (eds), *Survey of London* (Vol. 20, 1940), on British History online, www.british-history.ac.uk

C.L. Falkiner, '(Richard) William Steuart Trench 1808–1872', *Dictionary of National Biography*.

Roy Porter, *London: a social history* (Penguin, 2000).

'Minutes of the evidence of the select committee on Trafalgar Square, 1840'.

13 Wren's missing marvels
Felix Barker and Ralph Hyde, *London as it might have been* (John Murray, 1982).

Kerry Downes, 'Christopher Wren 1632–1723', *Dictionary of National Biography*.

Antonia Fraser, *King Charles II* (Weidenfeld & Nicholson, 1989).

Stephen Inwood, *A History of London* (Macmillan, 1998).

Adrian Tinniswood, *His invention so fertile: a life of Christopher Wren* (Jonathan Cape, 2001).

14 The tumbling abbey habit

John Rutter, *Delineations of Fonthill and its Abbey* (self-published 1823, facsimile published by Gregg Publishing, 1973).

Henry Venn Lansdown, 'Recollections of William Beckford: on Beckfordiana – a contemporary assessment by a friend of Beckford's', on the William Beckford Society website, www.beckford.c18.net

William Donaldson, *Brewer's Rogues, Villains and Eccentrics* (Weindenfeld & Nicholson, 2002).

E.A. Smith, *George IV* (Yale, 1999).

15 Why Lutyens' cathedral vanished

Liverpool Museums Publicity documents 2007.

Liverpool Metropolitan Cathedral website, www.liverpoolmetrocathedral.org.uk

Time Magazine (12 Aug 1929).

The Independent (9 July 1966).

Liverpool Daily Post (28 Nov 2006).

Apollo (1 Jan 2007).

16 New York's doomed dome

R.M. Marks, *The Dymaxion World of Buckminster Fuller* (Southern Illinois University Press, 1960).

H. Kenner, *Bucky: A guided tour of Buckminster Fuller* (William Morrow &Co., 1973).

The Buckminster Fuller Institute at Santa Barbara, www.buckminsterfuller.net

17 Exploding traffic lights

Westminster City Council leaflet and other materials from the Metropolitan Police Historical Archive, West London.

Daily Mail (5 March 1998).

'Minutes of the proceedings' (Institute of Civil Engineers, 1887).

Police Review (29 April 1983).

18 The steam-powered passenger carriage

Di Thomas Kingston Derry and Trevor Illtyd Williams, *A short history of technology: from the earliest times to AD 1900* (Oxford University Press, 1960).

Alain A. Cerf, 'Nicholas Cugnot, Fardier 1770', on www.nicolascugnot.com

Di Lita Epstein, Charles Jaco and Julianne C. Iwersen-Niemann, *The complete idiot's guide to the politics of oil* (Alpha Books, 2003).

Samuels Smiles, *The lives of engineers: George and Robert Stephenson* (Salzwasser-Verlag, 2010).

'The Stanley Steam Engine', on www.stanleymotorcarriage.com

19 Flying cars

Life (16 August 1937).

Popular Science (2000).

Times (15 May 2004).

The Seattle Times (5 Sep 2006).

The Seattle Times (15 July 1990).

New York Times (11 April 2009).

New Scientist (29 May 1999).

20 The atomic automobile

Original Publicity Documents, Ford (1958).

'When Dream Cars Collide With Real-World Demands', *New York Times* (7 January 2007).

'Nuclear Powered Passenger Aircraft to Transport Millions', *Times* (27 October 2008).

21 The X-ray shoe-fitting machine

Leon Lewis and Paul E. Caplan, 'The shoe-fitting fluoroscope as a radiation hazard', *California Medicine* (Vol. 72, No 1, January 1950).

Council A. Nedd II, 'When the Solution Is the Problem: A Brief History of the Shoe Fluoroscope', *American Journal of Roentgenology* (June 1992).

Charles Hartridge, (letter by), 'Shoe X-Ray machines', *New Scientist* (3 March 1960).

'Fluroscope a success', *New York Times* (12 May 1896).

Dr Michael J. Smullen and Dr David E. Bertler, 'Basal cell carcinoma of the sole: possible association with the shoe-fitting fluroscope', *Wisconsin Medical Journal* (2007).

22 The cure that killed

Paul W. Frame, *Radioactive curative devices and spa* (Oak Ridge Associated Universities, 1989).

Dr Mark Neuzil and Bill Kovarik, *Mass Media & Environmental Conflict* (Sage Publications, 1996).

Laura Sternick, 'Liquid Sunshine: The Discovery of Radium', *Journal of Science* (2008).

Nanny Fröman, 'Marie and Pierre Curie and the Discovery of Polonium and Radium', on www.nobelprize.org

'Health Consultation: US Radium Corporation', State of New Jersey
(June 1997).

'Poison Paintbrush', *Time Magazine* (4 June 1928).

'Radium Drinks', *Time Magazine* (11 April 1932).

'Make luminous drinks from radium', *New York Times* (14 January 1904).

23 The 'cloudbuster'

Mildred Edie Brady, 'The Strange Case Of Wilhelm Reich', *Bulletin Of The Menninger Clinic* (March 1948).

R.Z. Sheppard, 'A family Affair: review of *A book of dreams* by Peter Reich', *Time* (14 May 1973).

Wilheim Reich obituary, *Time* (18 Nov 1957).

Judge Clifford, 'Decree Of Injunction Order' (USA vs. Wilhelm Reich) (March 19, 1954).

Wilheim Reich, *Die Bione* (1938). English translation on: www.actlab. utexas.edu/~hahmad/bions.htm

24 The brand new continent of Atlantropa

Willy Ley, *Engineers Dreams* (1955).

Cabinet Magazine (Spring, 2003).

Gibraltar Magazine (January 2009).

The Independent (24 July 2004).

Star-Tribune (14 July 1999).

25 A nation built on sand

The Economist (24 December 2005).

Harper's Magazine, (October 2004).

Cabinet Magazine (Issue 18, 2005).

26 The Darien debacle

Spencer Collection, University of Glasgow.

The Guardian (11 Sep 2007).

The Scotsman (28 April 2007).

Dr Mike Ibeji, 'The Darien Venture', on www.bbc.co.uk

History Today (1 Nov 1998).

27 The lost US state of Transylvania

Christian G. Fritz, *American Sovereigns: The People and America's Constitutional Tradition Before the Civil War* (Cambridge University Press, 2008).

Michael J. Trinklein, *Lost States: True Stories of Texlahoma, Transylvania, and Other States That Never Made it* (Quirk Books, 2010).

The Columbia Encyclopaedia (Sixth Edition).
John E. Kleber, *The Kentucky Encyclopaedia* (1992).

28 The French republican calendar
David Ewing Duncan, *The Calendar* (Fourth Estate, 1998).
Matthew John Shaw, *Time and the French Revolution, 1789–Year XVI*
(M.Phil thesis, University of York, 2000).
'The Fondation Napoléon', on www.napoleon.org
'Calendars through the ages', on www.webexhibits.org/calendars

29 Latin Monetary Union
Kee-Hong Bae and Warren Bailey, *The Latin Monetary Union: some
evidence on Europe's failed common currency* (Korea University and
Cornell University, 2003).
Luca Einaudi, *European Monetary Unification and the International Gold
Standard (1865–1873)* (Oxford University Press, 2001).
Henry Parker Willis, *A history of the Latin Monetary Union: A study of
international monetary action* (University of Chicago Press, 1901).

30 A tax on light and air
Glyn Davies, *A history of money from ancient times to the present day*
(University of Wales Press, 2002).
Andrew E. Glantz, 'A Tax on Light and Air: Impact of the Window Duty
on Tax Administration and Architecture, 1696–1865', *Penn History
Review* (Vol. 15, Issue 2, University of Pennsylvania, 2008).
G. Timmins, 'The History of Longparish', on www.longparish.org.uk/
history/windowtax
W.R. Ward, 'Administration of the Window and Assessed Taxes 1696–
1798', *The English Historical Review* (Oxford, 1952).
Stephen Dowell, *A History of Taxation and Taxes in England* (Longman
Green, 1884).

31 Paxton's orbital shopping mall
Independent (5 May 2009).
Telegraph (29 June 2005).
Telegraph (12 August 2003).
Georg Kohlmaier, Barna von Sartory and John C. Harvey, *Houses of
Glass: A Nineteenth Century Building Type* (The MIT Press, 1991).
The Civil Engineer and Architect's Journal (Vol. 18, 1855).
James Winter, *London's Teeming Streets* (Routledge, 1993).
'Minutes of Evidence from the Select Committee on Metropolitan
Communications, 1855.'

32 Cincinnati's subway to nowhere
Allen J. Singer, *The Cincinnati Subway* (2003).
Cincinnati Enquirer (24 May 2003).
Cincinnati Enquirer (29 July 2002).
City of Cincinnati government website, http://www.cincinnati-oh.
    gov/
Cincinnati Examiner (1 June 2010).
Cincinnati Post (23 March 2007).

33 Is it a train or a plane?
'The George Bennie Railplane System', Brochure, National Archives of
    Scotland.
Christian Wolmar, *Blood, Iron & Gold* (2009).
William B. Black, *The Bennie Railplane* (East Dunbartonshire Council,
    2004).
The Times (13 March 2006).
The Guardian (13 March 2006).
Milngavie Herald (23 Aug 2006).
The Sunday Times (18 April 2010).

34 How the Cape To Cairo railway hit the buffers
Albert A. Hopkins, *Scientific American Reference Book* (A. Russell Bon,
    1905).
George Tabor, *The Cape To Cairo Railway and River Routes* (Genta
    Publications, 2003).
New York Times (8 March 1908).
J.G. Macdonald, *Rhodes – A Life*.

35 From Russia to America by train
James A. Oliver, *The Bering Strait Crossing* (2006).
Time (13 March 1944).
'Extreme Engineering', *Discovery Channel* (2003).
Popular Science Magazine (December, 2004).

36 British Rail's flying saucer & the 'great space elevator'
The Complete Dictionary of Scientific Biography.
Michel Van Pelt, *Space Tethers and Space Elevators* (Praxis Publishing,
    2009).
Daily Mail (28 Feb 1996).
The Guardian (29 April 2006).

37 Why the bathyscaphe was the depth of ambition

Bill Bryson, *A short history of nearly everything* (Doubleday, 2003).

T.A. Heppenheimer, 'To the bottom of the sea', *Invention and Technology Magazine* (Vol. 8, Issue 1: summer 1992).

Robert Gannon, 'What really happened to the Thresher', *Popular Science* (February 1964).

'The deepest explorers', a website dedicated to Auguste and Jacques Piccard: www.deepestdive.com

38 The perpetual motion machine

William H. Huffman, *Robert Fludd and the end of the Renaissance* (Routledge, 1988).

Arthur W.J.G. Ord-Hume, *Perpetual Motion: The History of an Obsession* (Adventures Unlimited Press, 1977).

Michael D. Lemonick, 'Will Someone Build A Perpetual Motion Machine?', *Time* (10 April 2000).

39 Brunel's not so *Great Eastern*

George S. Emmerson, 'L.T.C. Rolt and the Great Eastern Affair of Brunel versus Scott Russell', *Technology and Culture* (Vol. 21, No 4, The Johns Hopkins University Press, October 1980).

'Great Eastern: The Launch Of The "Leviathan"', *Mechanics Magazine* (19 December 1857).

Keith Hickman, 'Brunel's "Great Eastern" Steamship, The Launch Fiasco – An Investigation', *Gloucestershire Society for Industrial Archaeology Journal* (2005).

For pictures and more information about the SS *Great Eastern* see www.brunel200.com

40 Bessemer's anti-seasickness Ship

Sir Henry Bessemer, F.R.S. An Autobiography (1905).

Percy Hethrington Fitzgerald, *Short Works of Percy Hethrington Fitzgerald* (Biblio Bazaar, 2008).

41 Escape coffins for the mistakenly interred

Jan Bondson, *Buried Alive: The Terrifying History of Our Most Primal Fear* (W.W. Norton, 2002).

Bill Bryson, *At Home* (Doubleday, 2010).

Stephen B. Harris, 'The Society for the recovery of persons apparently dead', *Cryonics* (1990) (at the Alcor Life Extension Foundation).

42 Chadwick's miasma-terminating towers

Stephen Halliday, 'Death and miasma in Victorian London, an obstinate belief', *British Medical Journal* (22 December 2001).

Di Samuel Edward Finer, *The life and times of Sir Edwin Chadwick* (Methuen, 1952).

Vladimir Jankovic and Michael Hebbert, 'Hidden Climate Change: Urban Meteorology and the Scales of Real Weather', Paper delivered at the Royal Geographic Society-IRG annual conference, 2009.

Horace Joules, 'A preventative approach to common diseases of the lung', *British Medical Journal* (27 November 1954).

A. N. Wilson, *The Victorians* (Arrow Books, 2003).

43 Bentham's all-seeing panopticon

London's Bentham Project, University College: www.ucl.ac.uk/Bentham-Project

Internet Encyclopedia of Philosophy: www.iep.utm.edu/bentham

Simon Schama, *A history of Britain: Volume 3, The fate of Empire 1776–2000* (BBC, 2002).

A. N. Wilson, *The Victorians* (Arrow Books, 2003).

44 The self-cleaning house

Colin Davies, *The Prefabricated Home* (Reaktion Books, 2005).

Kristin McMurran, 'Frances Gabe's Self-Cleaning House could mean new rights of spring for housewives', *People Magazine* (29 March 1982).

Kim Elliott, 'Frances Gabe: At home in Futureville' (2009), on www.serendipityjones.com

'Vatican paper: washing machine liberated women most', *Reuters* (9 March 2009).

45 The jaw-dropping diet

Horace Fletcher, *The A.B.Z of Our Own Nutrition* (1903).

Horace Fletcher, *Fletcherism: What is it or how I became young at Sixty* (1913).

David F. Smith, (ed.), *Nutrition in Britain* (1997).

Daily Mail (14 March 2008).

New Scientist (6 Nov 1980).

New York Times (14 July 1919).

46 A nutty plan to feed the masses
History Today (1 Jan 2001).
Alan Wood, *The Groundnut Affair* (Bodley Head, 1950).
D.R. Myddelton, 'They Mean Well: Government Project Disasters', *The Institute of Economic Affairs* (2007).
The Independent (15 Jan 2010).

47 Ravenscar: the holiday resort that never was
The Northern Echo (19 Feb 2001).
Yorkshire Post (13 November 2004).

48 Why Smell-O-Vision stank
Mark Thomas, *Beyond Ballyhoo: Motion Picture Promotion and Gimmicks* (McGee, McFarland, 2001).
Scott Kirsner, *Inventing the Movies* (CinemaTech Books, 2008).
The Independent (10 May 2002).
New York Times (8 May 2002).
Martin J. Smith and Patrick J. Kiger, *OOPS: 20 Life Lessons From the Fiascoes That Shaped America* (Collins, 2007).
Daily Mail (21 Oct 2010).

49 The metal cricket bat
Martin Williamson, *Cricinfo Magazine* (25 Sep 2004).
Dennis Lillee, *Menace* (Hodder, 2003).
Australian Broadcasting Corporation (16 Dec 2003).

50 Bicycle polo & other lost Olympic sports
Bicycle Polo Association of Great Britain, History & Rules (1938).
The Guardian (23 August 2004).
The Scotsman (16 August 2008).
The Observer (1 August 2004).

Follow us on Facebook: Pigeon-Guided-Missiles

Visit our website and discover thousands of
other History Press books.

www.thehistorypress.co.uk